THE TRANSFER OF TECHNOLOGY TO SOCIALIST COUNTRIES

The Institute and its Research Committee are grateful for the comments and suggestions made by Michael Kaser and Dr Peter Marsh, who were asked to review the manuscript of this book.

The Transfer of Technology to Socialist Countries

The Case of the Soviet Chemical Industry

V. SOBESLAVSKY
P. BEAZLEY

Published for
THE ROYAL INSTITUTE OF INTERNATIONAL AFFAIRS
by
Gower

OG &H Oelgeschlager, Gunn & Hain, Publishers, Inc.
Cambridge, Massachusetts

Published by

Gower Publishing Company Limited,
Westmead, Farnborough, Hants., England

and

Oelgeschlager, Gunn & Hain, Publishers, Incorporated
1278 Massachusetts Avenue
Cambridge, Massachusetts 02138, USA

ISBN 0-566-00325-2
0-89946-040-2 (U.S.A.)

Contents

Abbreviations

CEFIC	European Chemical Manufacturers Federations (Centre Européen des Federations de l'Industrie Chemique)
CMEA	Council for Mutual Economic Assistance (also known as Comecon)
COCOM	Co-ordinating Committee for East-West Trade Policy
EEC	European Economic Community
GKNT	State Committee for Science and Technology ('*Gos'tekhnika*', also known as SCST)
NBEZ	Non-Black Earth Zone of the Russian Soviet Federative Socialist Republic
NEP	New Economic Policy

Chemical products abbreviations

DMT	dimethylterephthalate
HDPE	high density polyethylene
LDPE	low density polyethylene
PVC	polyvinylchloride
SBR	styrene butadiene synthetic rubber
SK-B	sodium butadiene synthetic rubber
VMC	vinylchloridemonomer

Introduction

The history of Soviet industrialisation is also a history of Soviet technological dependence on the outside world. This in itself would be nothing remarkable as international technology transfer is just as essential an ingredient of industrial development in the USSR as it has been in all the other industrialised countries in the world. Nevertheless, the case of the Soviet Union is special because after the 1917 revolution the USSR embarked on the road of constructing a new economic and social order which in every respect is an antithesis of Western capitalism. Moreover, since the construction of communism has been accompanied by economic autarky, political isolationism and cultural as well as spiritual separation, the Soviet readiness to benefit from the technological advances accomplished in the West constitutes a striking exception amidst this overwhelming tendency towards closeness and hostility in respect to the outside world.

The story of Soviet technological dependence on the West has an ironical twist. Its rationale has demanded that the Eastward flow of technology has as a rule been facilitated by large multinational companies who alone possess the necessary research and development (R and D) back-up, and managerial and financial resources of the magnitude and quality required by such a giant as the USSR, but who, for ideological reasons, have always been regarded by the latter as the very bastions of capitalism and imperialism whose might, from the Soviet point of view, should at all times and places be uncompromisingly opposed and curtailed. This flexibility in the realms of economic policy and ideology can be traced in all stages of the genealogy of Soviet communism. Soon after the revolution Lenin argued in favour of co-operation with interested Western firms. Despite some opposition from his colleagues, his views prevailed and a number of companies were given the rights, or 'concessions' as they became known, to develop and exploit the Soviet Union's raw materials potential, thus speeding up the industrialisation of their socialist partner. During the years of Stalin's rule the policy of 'concessions' was gradually halted but the USSR continued to benefit from the acquisitions of foreign technology which were secured by a variety of means ranging from duplication of purchased prototypes and monitoring of foreign scientific literature to direct Western assistance during the war against

Nazi Germany. The Soviet Russian policy of 'concessions' and, more recently, of a far more comprehensive industrial co-operation has from its early days contained an element of tactical compromise with foreign capitalism which all the Soviet leaders, from Lenin to Brezhnev, found perfectly consistent with their professed long-term strategy of assisting the ultimate collapse of the latter by overt and covert means. (The strengthening of the Soviet economy through co-operation with Western firms, far from representing a retreat, can with a little or no ideological difficulty be interpreted in terms of fortifying the socialist mother-land and speeding-up of the process of the world-wide revolution.) Thus the East-West political and commercial relations have always been conducted amidst an atmosphere of caution and outright hostility as well as enthusiasm and often unjustified optimism about the reduction of tension and increase of business opportunities.

After the death of Stalin the Soviet leaders and economic planners grew more acutely aware of the large and ever-increasing technological gap that separated their country from the rest of the industrialised world, and they responded by initiating an impressive programme of rapid modernisation of the economy which was finally endorsed in 1958 by the launching of the Seven Year Plan. This programme was accompanied by gradual termination of the Cold War between the East and West and by the dawning of the era of peaceful co-existence and subsequent *détente*. As in the earlier stages of Soviet industrialisation, the programme of modernisation started by Khrushchev and accelerated and further deepened under the leadership of Brezhnev would be immensely harder to implement were it not for the Soviet policy of importing advanced technology from the West.

A part and parcel of the 1958 plan was an expansion and modernisation of the chemical and petrochemical industries. The success of the 'chemicalisation' drive, as this programme became known, was particularly heavily dependent on Soviet imports of complete chemical plants from the West and on purchases of manufacturing licences, because these industries are among the most research intensive industrial branches; moreover, the Soviet lack of R and D expertise in this area was especially pronounced owing to the excessive concentration on coal at the expense of oil products traditionally favoured under Stalin. Today, after twenty years of 'chemicalisation', the expansion of the industry in the USSR continues unabated and technological co-operation with Western multinationals has, if anything, increased in scope and become more varied in its forms.

The present study addresses itself to this particular aspect of West-East transfer of technology. By examining in detail the structure of the Soviet chemical and petrochemical industries as well as the intentions of the Soviet planners, the study illustrates the role performed by imports of Western plants and equipment in realising the tasks of 'chemicalisation'. The study, then, is a case study of East-West transfer of technology in a limited, but relatively very important area, which means that the issues raised in it in respect to 'chemicalisation' in the USSR have much wider implications in the realm of economic and political relations between the socialist countries of Eastern Europe and the West. Our aim is to draw on the results of research into the subject undertaken to date, but also to supplement it by new information and conclusions reached by us and by a group of experts who regularly convened at Chatham House, and thus to present a contribution to the current debate on the economic and political implications of East-West technology transfer.

Since we believe that the main reason why the Soviet Union (and the other socialist countries) has been compelled to rely on capitalist technological dynamism and resourcefulness should be sought in its rejection of market mechanism as the basis on which its economy operates, Part I of the study is devoted to a discussion of the main features of economic management of a Soviet-type economy and to an illustration of how these features inhibit the generation and diffusion of technological progress throughout the economy. Thus chapter 1 outlines how the Soviet system functions and in addition it discusses some of the most important changes in the political and economic climate under which this system operates, thereby bringing into a fuller relief the essence of the Soviet current economic predicament. In this way we hope not only to throw additional light on the motives that lie behind the Soviet quest for foreign technology, but also to explain why in recent years this quest has become more intense than ever before. Chapter 2 continues by evaluating Soviet performance in the area of technological innovation, and it takes up the preceding discussion of the system's characteristics by focussing in greater detail on the concrete causes of the sluggish rate of technological change produced by the essential systemic features outlined earlier. Part II constitutes the empirical core of the study: chapter 3 looks at the structure and development of the Soviet inorganic and petrochemical industries and the role of Western technology within them. Chapter 4 examines the Soviet chemical fertilisers industry in the light of the overall agricultural policy pursued by the leadership, and like the preceding chapter it illustrates the role of imported technology in its build-up. Chapter 5 discusses the emergence of a relatively novel form of East-West commercial co-operation – the countertrade (compensation or 'buy-back') agreements – with special reference to its likely implications in the area of chemical production. Finally, chapter 6 assesses the importance of East-West transfer of technology in terms of both political and economic relations between the East and West by drawing together a number of major issues discussed earlier.

A few explanatory words are necessary at the outset in regard to the overall approach to the phenomenon of East-West transfer of technology as adopted in this study and as briefly outlined above. This approach starts from the premise that East-West technological co-operation and commercial relations generally cannot be discussed in the same manner as one would treat the subject of technological flows between, for example, the United States and Western Europe or Western Europe and the developing nations. The East-West technology transfer takes place between countries who not only are at disparate levels of development but also possess vastly different political and economic systems. However, even this statement falls short of fully capturing the special property of East-West relations in general and East-West technology transfer in particular. In addition we have to bear in mind that these two political and economic systems have for decades been engaged in a permanent state of rivalry and hostility as well as in accommodation and industrial co-operation. We may even go further and suggest that, despite the hostile rhetoric, the Soviet leaders are aware that they, and their socialist centrally planned system, have a vested interest in a continued prosperity of the Western world because there are areas – and the generation of technological progress is the foremost of them – in which reliance on purely domestic resources would lead to a blind alley.

In order to capture this ambivalent essence of East-West relations and in order that its presence can be traced in the specific subject of East-West transfer of technology, the present study relies heavily on its method of apposition of the concepts of (free) market and centrally planned economy. Despite certain limitations and dangers of over-simplification implied in our approach, we are of the opinion that the process of East-West technology transfer itself, as well as its causes and consequences, could not be more accurately described and analysed in a way which would completely fail to refer to such apposition. Rightly or wrongly, for the purposes of our argument, we assume that there is a causal link between a given country's system of economic management and its performance in the area of technological progress. This means that if country A wishes to terminate its chronic technological dependence on an industrial, and hence technological, leader B it would be well advised to begin such efforts by attempting to adopt B's system of economic organisation or to find some other means by which it could approximate to the conditions generated by B's system. Our next assumption is that, viewed from the point of the market-plan scale, most of the industrially developed countries of the West may be said to have succeeded in finding an optimum balance between the imperatives of state intervention on the one hand and the necessity of preserving the freedom of the market and private initiative on the other. (This optimum 'mixture' blended by the mixed economies of the industrially advanced Western countries cannot, of course, be seen in isolation from the prevailing practice of political pluralism which, by virtue of its ability to accommodate a variety of conflicting interests, produces a parallel 'mixture' in the domain of politics.) Now, any country wishing to achieve a comparable degree of excellence in the areas of science and technology (and its application along the broad front of economic activity) should start by duplicating, at least in some respects, the economic systems of the advanced Western countries.

In this scenario the Soviet Union (and the other socialist countries) represents the country A and the industrially advanced West is the country B. The complete (or near complete) exclusion of market-based tools from playing their part in the process of economic direction (and the ban on political pluralism) instituted by the former has led to a situation of imbalance (or absence of optimum) in the market-plan relation. This imbalance is in its turn responsible for the propensity to co-operate with the West, despite the vehement Soviet disapproval of capitalism, because such co-operation offers an opportunity for a partial restoration of balance while the domestic arrangement need not be reformed. For political reasons it is perhaps impossible in the foreseeable future to transform the Soviet system in such a way as to make it like that of the advanced Western countries but there are no reasons why the most valuable product of these countries – their advanced technology and superior managerial methods – should not be imported ready-made. To put it in another way, in importing advanced technology from the West, the East may be said to be engaging in a policy of importation of the optimal market-plan 'mixture' which it is unable and, indeed, unwilling to produce domestically. This, then, is the core of our argument, and the layout of this study is organised in a manner that unfolds it with maximum efficacy.

The reader is, therefore, warned that our frequent references to the virtues of free market and failures of strict central planning are not meant to amount to unqualified praise of the principles of *laissez-faire*. Our aim has been to develop a coherent theoretical framework into which our discussion of the complex subject of East-West transfer of technology could be placed and this task, for the reasons outlined above, implies the assumption that, at least in the area of science and technology (and, especially, of its application) the market-based systems have so far performed better than systems based on strict central planning. In setting this aim we wish to contribute to the debate on the wide implications of East-West transfer of technology and East-West relations in general. By the same token, the general subject of weighting the vices and virtues of centrally planned systems against those of market-based ones should not be regarded as our main preoccupation. In any case, such a subject could not be satisfactorily dealt with in a monograph of the dimensions of the present one.

Finally, a brief note regarding the terminology used in the present discussion. Since the flow of resources we are analysing refers to an *Eastward* movement it would, technically speaking, be correct to refer to it as West-East technology transfer. However, such a term would go against the grain of the generally accepted notion of East-West relations of which West-East transfer of technology is an intrinsic part. For this reason we shall continue to use the term East-West, with the provision that strictly speaking it denotes a process of West-to-East transfer of technology.

We wish to acknowledge our indebtedness to the members of the Chatham House study group and to a number of other people who have assisted us in this study. In particular we have drawn upon the wealth of information available in the literature published by Soviet specialists in the academic field, on technical journalists, members of the chemical industry, contractors, bankers, members of official bodies and many others. We especially acknowledge the value of the work of Mr A.C. Sutton, *Western Technology and Soviet Economic Development* (Hoover Institution Press, Stanford), Dr R. Amann's contribution on the Soviet chemical industry in *Technological Levels of Soviet Industry* (Yale University Press) and Professor J. Berliner's study *The Innovation Decision in Soviet Industry* (MIT Press). The Compendium of Papers comprising *The Soviet Economy in a New Perspective* submitted to the Joint Economic Committee of Congress of the United States was also widely drawn upon. *The European Chemical News*, *Chemical Age* and *Chemical Insight* were all invaluable in their reporting of current developments and interpretation of latest trends. The permission granted by Chemical Data to utilise their maps as the basis of the maps produced for the present study is also greatly appreciated. To these and all others who have given advice and criticism we are much indebted. The opinions expressed in the study and the errors we may have committed are our own.

V. SOBESLAVSKY
P. BEAZLEY

PART I

1 The Soviet economic system: its strength and limitations

A Free market *v.* central planning: a historical overview

The historical evolution of the Soviet economic system may be most profitably analysed in terms of the system's ambivalent relationship with the concept of free market economy. Free market, being the essence of capitalism, was not simply rejected by the founders of socialism in Soviet Russia; its suppression constitutes, above all, the legitimation and *raison d'être* of the subsequent application of strict central control over economic activity in its totality. However, since market is one of mankind's most ancient institutions, the Bolsheviks' success in their determined efforts to eliminate it for good was bound to be limited. Depending on the economic strength and confidence of the leadership, the role of the market-element in the economy's functioning either increased or diminished in the course of the Soviet system's history.

When Soviet Russia embarked upon the road of socialist construction it was an industrially backward country, ravaged by the years of war and revolutionary upheavals which came in its wake. From the 1917 October revolution until 1921 the economy was fully nationalised and run on the principles of strict centralism and tight control over the limited resources and productive capacity that still operated. The foodstuffs supplies were secured by the means of forced requisitions of what the peasants managed to produce, and under such conditions normal economic relations, including the use of money, ceased to apply. This period, often referred to as war communism, came to its end when Lenin realised that the peacetime task of restoring prewar levels of production could not be accomplished by these drastic means. Much against the wishes of many Bolsheviks he called off the war communism and initiated his New Economic Policy (NEP), which was to last until 1928-29 when Stalin launched his rapid industrialisation drive.

NEP represented a retreat towards the concept of free market. Its essence was a partial de-nationalisation of the means of production; large scale industry, banking and foreign trade remained in the hands of the state but small traders and manufacturers were given a more or less free hand in conducting their affairs. The peasants were encouraged to produce for the market and to get rich in the process while industrial managers, civil servants and other professionals returned to their jobs. The state enterprises themselves proceeded to operate within the paradigm of the re-emerged market rather than in accordance with the principles of central planning. Foreign capital was allowed to enter the Soviet Union in the form of concessions given mainly in the industries extracting and processing raw materials. By 1925 NEP reached its high point. Growth was rapid and the new policy seemed to be succeeding beyond reasonable expectations, although this was due mainly to re-activating of existing capacity and re-absorption of available labour.[1]

The moral of the NEP period in the history of the USSR is central to purposes of the present study. It demonstrates that the Soviet leaders and the system they represent are capable of combining, in a very remarkable way, pragmatism and flexibility on the one hand with extreme inflexibility on the other hand. It may, in fact, be concluded that the temporary compromise of NEP was a tactical retreat necessary in order to survive in the face of unanticipated odds (indeed, it was at the time aptly described as the economic Brest-Litovsk) and thus to preserve the inflexible ideological postulates of communism. Throughout NEP the Party's grip on the 'commanding heights' of the economy never weakened, and when the new policy ran out of steam and decisions had to be taken as to what strategy to follow next the leadership, dominated then by Stalin, embarked upon the course of breakneck industrialisation and collectivisation, the speed of which exceeded anything that was even contemplated before. The system that evolved in response to the needs of this 'revolution from above' may thus be regarded as the restoration of the principles of strict central planning at the expense of free market. The centrally determined and implemented plan became the pivotal point around which everything else revolved. The market was squeezed out, never to return to the USSR again in the role it played during NEP. Capitalism, however, continued to play an important, if not the key, role in the further development of the Soviet Union. As this study shows, the USSR continued to rely on the technological achievements of the market-type economies and will have to do so for so long as its economy is controlled exclusively by strict central planning or, at any rate, for so long as it fails to adapt its system to the vastly changed environment of today.

B The main features of the system

The Soviet economic system can be compared to a giant corporation which monopolises the whole totality of economic activity within the country.[2] It makes sure that it has the necessary capital stock, it sells its consumer goods and services and it charges what prices it considers necessary for both its consumers' and producers' goods. It decides what type and what rate of taxes it will impose and where it will incur loss or make a profit. It controls foreign trade by deciding upon the types of goods and quantities to be exported and imported, and it subsidises any losses made in the process. This system came to embody the general understanding of what socialist economics is about in spite of the fact that it has undergone many changes and administrative reorganisations. This is so because its fundamental property, strict central planning as the guiding principle, has not been affected. Some of the East European socialist countries who in the 1950s adopted the Soviet-type system, notably Hungary, are today so far from this *ancien régime* that it would be misleading dogmatically to hold on to this concept as the universal model of socialist economics. (The basic information about the international economic organisation of the socialist countries – the Council of Mutual Economic Assistance – is given in Appendix A.) Nevertheless, the useful purpose this description serves is to set ourselves a clearly delineated starting point for the subsequent discussion. It can be used as a suitable bench-mark for the purposes of the present study.

The basic unit of the centrally planned system is the enterprise whose main task is to fulfil the plan handed down to it by its ministry. Depending on the type of the industry in question, there may be intermediate administrative bodies between the ministry and the enterprise, chief administration or *glavki* in the Soviet Union, but these have recently been reorganised as part of the policy of forming larger industrial units, the so-called production or industrial associations.

The apex of the ministerial system is the All-Union Council of Ministers. This body is responsible for plan fulfilment and overall performance of the economy to the political masters of the Politbureau of the Communist Party. Within the Politbureau itself, responsibilities are further divided so that individual members of the Politbureau may supervise a particular sector of the economy (or some other facet of government).

The process of planning is a complex one with its main features being the task of drawing up material balances necessary for ensuring that a certain quantity of a given product can be produced and that appropriate products meet with their utilisation.[3] The most important bodies engaged in the planning process are the State Planning Committee, *Gosplan*, and the State Supply Committee, *Gossnab*. *Gosplan*, supported by an impressive organisational structure, research institutes and computer centre ensures an overall coherence in drawing up the main 2,000 material balances. *Gossnab* acts as an advisory body to the

Council of Ministers and *Gosplan* and it looks after the balances of about 18,000 less important products.[4] The priorities in allocation of resources are determined by the Politbureau but, obviously, this involves some tough bargaining since the conditions of taut planning produced by the system render all scarcities of resources especially acute. The Politbureau members eventually hammer out the order of priorities and the resources are distributed accordingly. Branches with highest priority get everything needed, and those with least priority get little or nothing.[5]

The plan itself is a document carrying the force of law. It tells the enterprise what to produce, where to get the inputs from and what to do with the output. It determines the pattern of financial flows by stipulating the amounts of payments into and out of the state budget. It specifies the rates of investment, profit and capital depreciation, the size of the wage fund, labour productivity and other matters. There are medium term plans for five years and annual operational plans with the highest degree of disaggregation. In addition the planners draw up long-term perspective plans for fifteen years or more. A limited amount of feedback does take place throughout the planning process because a fundamental component of the procedure is the assembling of draft plans by the enterprises themselves. These drafts are formed in accordance with a broad set of directives issued by the central planners and the centre, having received the drafts, co-ordinates them and sends them down as final instructions to the enterprises.[6]

Since market, as pointed out above, has been pushed out into playing no or a very limited role in determining the behaviour of enterprises and their supervisory bodies, prices have ceased to perform their role as carriers of information and determinants of the movement of resources. Prices are used primarily for the purposes of accounting, and this implies that their role *vis-à-vis* the macro- and micro-economic decision-making is a passive one. Accordingly, profit has also ceased to be the main indicator of an enterprise's success or failure. Instead it is the degree of plan-fulfilment that is decisive. The Soviet economy, it must be remembered, operates within the framework of its own social, political and ideological values: hence its disregard for the traditional ones.

The application in practice of these principles generates a whole spectrum of other peculiarities which, depending on one's approach, may be viewed either as logical consequences of the adoption of central planning, or as integral parts of the preconceived master plan constructed by the designers of the system. We shall discuss only a few of the most important characteristics.

C Other important features

Politicisation of economic life

It is notoriously difficult to disentangle economic from political considerations, but within the socialist paradigm this task is twice as hard. The fundamental difference between the Western (i.e. non-communist) concept of socialism and the socialist countries' practice of it has been the variance in the degree of militancy against capitalism. The Bolshevik and the communist parties are, after all, the heirs of the militant, non-reformist, tradition within the socialist movement. This seems to be the cause of the tendency now deeply ingrained in the system towards approaching its tasks in the spirit of the relentless revolutionary struggle. The whole economic, as well as any other, activity is supervised at every level by the organisational gamut of the Party, and in this way the Party becomes enmeshed in the process of economic decision-making and takes responsibility for any outcome, be it a success or a failure. Although material incentives are increasingly relied on in efforts at increasing output or quality, the workers are constantly encouraged by means of so-called moral incentives – mostly in the form of 'socialist competition' among factories or individuals for the attainment or overfulfilment of a given task. The 'socialist competition' is sponsored and supervised by Party organisations, and thus production frequently becomes an intensely politicised process. From the macroeconomic point of view, the politicisation of economic life is simply an expression of the fact that all means of production are nationalised and controlled by the state and ultimately by the Party.

Autarky and bilateralism

The industrialisation launched by Stalin, therefore, bore the hallmarks of a revolutionary campaign, very much in the tradition of the militant socialist party, rather than of peacetime construction policy. Moreover, it was accomplished against the background of 'capitalist encirclement' in the spirit of heroic struggle for building 'socialism in one country'. These external, as well as internal, dimensions contributed to the fostering of autarkic tendencies and attitudes. The Soviet Union possessed vast natural and human resources which needed to be developed and tapped and there was, consequently, little need and little will for engaging in large-scale foreign trade activities. Foreign trade came to be wholly subordinated to the exigencies of domestic planning which in practical terms means that, despite some evaluation of comparative costs, imports usually represent products that cannot be turned out domestically. The spendings on imports are then recouped by exports.[7] The role of foreign trade, therefore, tends to be passive. This practice leads to bilateralism because bilateral relations are a logical extension of central planning into the sphere of foreign trade. Bilateralism is a suitable tool by which foreign trade flows can be controlled in accordance with the requirements of the plan. It eliminates undesirable effects of price fluctuations and thus it effectively isolates the economy from essentially alien external influences. Foreign trade is conducted by Foreign Trade Organisations who act as monopolistic buyers and sellers of a particular type of commodity, with the result that enterprises themselves do not directly relate to their foreign suppliers or customers. At the same time, these peculiarities of the system do not prevent the socialist countries from playing a more active part, if they wish to do so, in international economic relations precisely because the domestic order is so effectively cut off from the external one.[8]

But bilateralism and autarky are also very deeply entrenched in the micro-economics of the system. This is because the central planning cannot in practice assure smooth supplies of necessary inputs throughout the economy, and ministries and enterprises react by creating their own capacities for the production of required inputs or agencies able to provide the required service. This leads to a considerable degree of parallelism, empire-building and duplication. Another consequence is the tendency to hoard raw materials and labour in order to cushion the enterprise in the event of a supply breakdown or when unexpected demands are made on it by the authorities.[9]

Curtailment of consumption

The Soviet Union started to develop its industrial base comparatively late. It had to catch up with the advanced Western countries in a relatively short period of time. Thus it happened that although, according to the theory prevailing in the socialist countries, the communist order will eventually bring about an abundance of life's necessities, for the time being it has been necessary to limit consumption as much as politically feasible and this, given the tight political control over the population exercised by the leadership, has meant that the rate of investment has traditionally been very high (25-30 per cent or more of national income). Indeed, one of the fundamental precepts of Stalinist economics was that investment should be as high as possible.[10] A corollary of this has been the precept that investment should be concentrated on sectors which produce the means of production and that these industries should at all times grow faster than the light and consumer-oriented ones. The Soviet economic development has, therefore, been lop-sided,[11] with disproportionate emphasis on the production of steel, energy, raw materials and heavy engineering and on the provision of an adequate network of heavy industrial transport facilities. This approach also accords with the traditional policy of assigning high priority to defence and defence-related industries. (Currently about 11-13 per cent of thc Soviet GNP is spent on defence, but the defence burden is much higher in the USSR than in the United States because, while there is a rough parity in the defence effort in both countries, the Soviet GNP is only about 50 per cent of the US GNP.[12])

As a result, emphasis has been placed on economic development on a national scale rather than on private luxury; on *economic* rather than *cultural* development; and on multiplying productive objects rather than things of beauty.[13] In such a situation the consumer is left with little if any sovereignty. The consumer market that does exist is a heavily distorted sellers' market. The socialist countries are well known for their queues in the shops and for the general lack of high quality goods and services. The curtailed consumer autonomy thus neatly synchronises with the limited political autonomy of the citizens.

Lop-sidedness

The lop-sided nature of the development has left a deep and very pervasive imprint on the economic and political life of the Soviet Union and the other socialist countries. The lop-sidedness might explain the extraordinary ability of these countries to excel in one respect while performing very disappointingly in another. Thus the USSR managed to achieve very high rates of growth and to become one of the world's superpowers in a relatively short period, but it finds it hard significantly to improve the quality of life of its citizens. The system has the ability to concentrate all available resources towards the attainment of one particular goal, but by the same token it must, of necessity, do so at the expense of other aims. In the domain of politics the lop-sided nature manifests itself *inter alia* in the fact that, although the rule of the Party is well entrenched, the authorities fear any expression of individual dissent on the part of a virtually insignificant section of the population.

The lop-sidedness also means that the apparent economic and political stability so prominently displayed on the surface tends to gloss over the weaknesses and instability discernible underneath. Perhaps the most significant consequence of this property of the Soviet-type system is that it finds it very hard, if not impossible, adequately to adjust to the changing economic, social and political conditions under which it operates. The most important weakness of the Soviet economic arrangement is that it is unable to meet the requirements of the increasingly more important intensive strategy of economic growth. The basic proposition frequently encountered in the literature on socialist economics is that the Soviet-type system was designed with a particular strategy of economic growth in view. That strategy is described as extensive, or classical, strategy of growth because, essentially, it is based on straightforward increase of inputs without paying attention to their quality and, by implication, to increased productivity. In the Soviet Union and the other socialist countries who adopted the Soviet model of industrialisation, this strategy took the form of excessive increase in low-productivity inputs from the countryside accompanied by an excessive rate of investment. The strict central direction of the movement of resources may be said to have suited the requirements of that strategy and this is corroborated by the fact that the USSR, and the other socialist countries, achieved very high rates of growth and became industrialised nations within a couple of decades. However, the drawback of this success is that it gradually turned into an obstacle to further progress, because the high growth rates of the past decades tend to be evoked as the ultimate proof of the essential soundness of the system, and consequently any reform of its structure becomes very hard to implement. But high rates cannot be indicative of economic progress when, like a Pyrrhic victory, cost exceeds benefits.[14] Maximum growth is not necessarily optimum growth, and as the pool of available labour resources in the countryside begins to dry up, and as the extraction of fuels and raw materials becomes more and more capital intensive and technologically demanding, the economy has to rely increasingly on the application of advanced technology, more sophisticated managerial methods and reduction of waste and inefficiency in general. The cost of achieving the high growth rates of the past now has to be paid back in terms of the system's unresponsiveness to the demands of our technological and consumer-oriented age. Great energy will have to be expended in order gradually to diminish the lop-sided nature of the system.

In chapter 2 we shall discuss some of the most important reasons why the Soviet economy cannot promote technological change at a satisfactory rate and hence why it has to rely on Western assistance in alleviating the effects of this incapacity. Before doing so, however, it might be useful briefly to outline the nature of the problems faced by those in charge of the Soviet economy in order to underline the existence of the factors that have acted as the main determinants of the intensity of the contemporary Eastward flow of technology.

D The Soviet economy and the intensive strategy of growth

Since the late 1950s, when the high rates of growth first began to show signs of the decline that proved to be a long-term trend, it has increasingly become apparent to Soviet economists and politicians that the existing tools for directing the economy were no longer adequate. There were basically three conditions which made the Stalinist economic model a viable proposition during the 1930-1950 period. The first was the above-mentioned availability of labour. The Soviet countryside provided a virtually limitless, as it then seemed, source of labour, and the relatively low priority assigned to agriculture itself meant that there were no obstacles to the policy of directing these resources into industry and construction. The second condition was the relative plenitude of raw materials necessary for this phase of Soviet industrialisation, and their extraction and transportation was not as capital intensive and technologically demanding as it is today. The third condition might best be described as the existence of Stalin himself. Under the atmosphere of terror it was relatively easy to maintain labour discipline by coercive means and it was possible to uphold the necessary high rates of investment and correspondingly low levels of consumption in order to sustain rapid industrialisation. None of these three conditions can any longer be taken for granted. Today it is recognised by both Soviet and Western economists, that the USSR is being increasingly constrained by actual or impending shortages in supply of the labour force[15] and raw materials, especially on the energy front.[16] As regards the third condition it too may be said to be departing gradually from the scene. Labour discipline has to be increasingly promoted by material incentives and this goes hand in hand with the increased consumerist expectations of the population. Consumption in its turn acts as a brake on the rate of investment, with the result that it too had to slow down during the current Tenth Five Year Plan (1976-1980).

Furthermore, over the years both Soviet and Western analysts have come to a general consensus that the decline in the rates of growth has not been a temporary phenomenon and consequently that there are deep-seated structural causes of this disappointing performance. To be sure, the rates of growth achieved by the Soviet Union, while the retardation was becoming more and more pronounced, were still very respectable by comparison with the developed Western countries (until the mid-1970s)[17] but the major source of concern has been the fact that the slow-down in the growth of output has occurred without a corresponding decrease in the rates of growth of inputs (see table 1). While the explanation of this phenomenon as well as of the registered slow-down of the rates of growth may be a subject of controversy, it

Table 1
Average annual rate of growth of GNP, inputs and productivity

	1951-60	1961-70	1971-75	1976-1980*
GNP	5.8	5.1	3.7	5.0
Inputs	4.6	4.3	4.3	3.5
Manhours	1.3	1.8	1.9	1.5
Capital	9.4	8.1	7.9	6.5
Land	2.4	0.4	0.8	0.5
Factor productivity	1.2	0.8	−0.6	1.5

*Planned

Source: CIA, *Soviet Economic Problems and Prospects* (Washington, 1977), p.10

seems that there is little doubt that one facet of the problem lies in the insufficient ability of the system to generate economic growth by means other than increases of combined inputs. There has generally been some factor productivity growth in the USSR, but to prevent retardation factor productivity would have not simply to grow, but to accelerate.

In theory, a large part of the *residual* of the growth of output that cannot be accounted for by increases of inputs is explicable in terms of technological change in the sense of introduction and diffusion throughout the economy of new products and processes. It is, therefore, obvious that the Soviet Union increasingly finds itself in a situation in which its performance in the area of introducing and successfully diffusing technological change becomes crucial in the task of attaining the rates of growth that the political leadership finds desirable. Although the rates of increase of inputs do not lose their significance, it is the rate of technological change and the rate of diffusion of technological change which gain salience.

Many observers of the Soviet scene would agree that the insufficient ability of the system to generate satisfactory rates of technological change and diffusion constitutes a major problem which is unlikely to be solved successfully in the near-term future. The publicity given to this issue in the Soviet Union and other East European countries indicates that the leaders are well aware of the shortcomings on this front. This does not necessarily imply that they harbour any hopes of returning to the period of high growth rates achieved during the early period of their industrial development. There is a general recognition of the fact that modern economy relying more and more on the intensive strategy of growth, cannot grow as fast as it could do at the start of industrialisation.[18] The more complex economy of today should, however, grow in its complexity. Figuratively speaking, we may say that it should expand in all directions rather than merely in the 'vertical' direction of expanding the physical volume of key items of heavy industry. It should be capable of delivering an increasing quantity of consumer goods of varying kind. It should aim, in other words, at improving the quality of life, and there are some signs that this has in fact, to a limited extent, been taking place. These dimensions of economic growth are not measurable simply in terms of industrial (or agricultural) output. And this is where the Soviet consumer enters the picture, for the leadership is increasingly forced to take into account the degree of consumers' satisfaction (or dissatisfaction) in the process of planning, setting the order of priorities, and forming their long-term policies. This imposes considerable political constraints upon the leadership which in itself is a remarkable and novel circumstance in the context of the Soviet economic and political environment. The retreat from the old attitudes and methods of evaluation has, however, been very slow. The Ninth Five Year Plan (1971-1976) was declared to be a consumer-oriented plan at the time when it was launched and for some time it seemed that a major re-thinking of priorities took place in the USSR because it was the first plan in the country's history which foresaw a higher growth rate of consumer industries than that of heavy industry. However, in the event the basic law of Stalinist economics remained unbroken; the consumer sector plan was not fulfilled and the system remained lop-sided. (Some observers may even be credited with guessing in advance that this indeed would be the case.)[19]

Notes

1 A. Nove, *An Economic History of the USSR*, Penguin, 1975, p.117.
2 P. Wiles, *Communist International Economics*, Basil Blackwell, Oxford, 1968, p.30.
3 A. Nove, *The Soviet Economic System*, Allen and Unwin, London, 1977, p.37.
4 Ibid., p.39.
5 I. Birman, 'From the Achieved Level', *Soviet Studies*, no.2, 1978, p.164.
6 It has been argued by Berliner that this in fact gives the enterprises more autonomy than is normally acknowledged. See J. Berliner, *The Innovation Decision in Soviet Industry*, MIT Press, 1976, p.15.
7 Nove, *The Soviet Economic System*, p.268.
8 *Banking, Money and Credit in Eastern Europe*, NATO Colloquium 1973, Summary, Brussels, 1973, p.11.

9 The enterprises are, for example, often required to help during the harvest the collective farms with whom they liaise.
10 Nove, *The Soviet Economic System*, p.155.
11 R. Hutchings, *Soviet Economic Development*, Basil Blackwell, Oxford, 1971, p.290.
12 H. Block, 'Soviet Economic Power Growth – Achievements under Handicaps' in *Soviet Economy in a New Perspective*, JEC, US Government Printing Office, Washington, 1976, p.267.
13 Hutchings, op.cit. (above note 11), p.291. Emphasis in the original.
14 J. Wilczynski, *Socialist Economic Development and Reforms*, Macmillan, London, 1972, p.24.
15 M. Feshbach and S. Rapawy, 'Soviet Population and Manpower Trends and Policies' in *Soviet Economy in a New Perspective*, JEC, 1976, p.113.
16 This is often due not so much to the USSR's not possessing adequate raw materials deposits as to the problem of extracting and transporting them to the consumption centres. The most notorious example in this respect is the case of oil. See, for example, L. Dienes, 'The Soviet Union: an Energy Crunch Ahead?', *Problems of Communism*, September-October 1977, p.46. A similar problem is faced by the USSR in regard to its labour force inputs: those nationalities which are registering relatively high birth rates (i.e. the Central Asian nationalities) have unfortunately traditionally displayed the lowest propensity to migrate to industrial centres where the labour supply shortages will be felt most. On this see Feshbach and Rapawy, op.cit. (previous note), p.124.
17 The results of the first two years of the current Five Year Plan indicate that the Soviet Union has registered the lowest rate of growth since World War II.
18 See, for example, the series of articles in *Pravda* in November 1977 by D. Valovoi.
19 K. Bush, 'The Tenth Five Year Plan and the USSR's Economic Prospects' in *Soviet Science and Technology: Domestic and Foreign Perspectives*, J.R. Thomas and U.M. Kruse-Vaucienne (eds), George Washington University, Washington D.C., 1977, p.265.

2 The Soviet system and technological progress

The preceding discussion illustrates the nature of the new demands produced by changed conditions under which the Soviet economic system operates and this, we hope, points to the growing importance of technological progress for a system that finds it difficult to respond with a satisfactory degree of flexibility. The Soviet Union has so far failed fundamentally to reform the basic principles of its economic organisation, but it did turn to the Western companies with offers of attractive business prospects in exchange for supplies of the most up-to-date technology. Part II of this study examines these technological relations in the relatively narrow but very important sector of the chemical and petrochemical industries. This chapter establishes the link between chapter 1 and Part II by presenting an evaluation of Soviet technological strength and by analysing in greater detail the main causes of Soviet shortcomings in the area of generation and diffusion of technological change.

A Soviet technology: the claims and achievements assessment

It has been observed that virtually all nations like to think that among other fields of human endeavour they also excel in the area of technology and that no nation is willing indefinitely to occupy the position of a technological follower dependent on advances accomplished abroad. This human passion of *technological chauvinism*[1] is, therefore, very widespread and no nation is totally immune from it. The Soviet Union and the Russian nation are no exception. The manifestation of *technological chauvinism* reached its peak there at the height of Stalin's rule when everything ever invented was attributed to the efforts of Soviet or pre-revolutionary Russian science. Under such conditions it became impossible objectively to assess the achievements of Soviet science and to compare it with the scientific and technological strength of the Western world. Moreover, in the USSR the notion of *technological chauvinism* gradually acquired an additional dimension, namely, that of *systemic (technological) chauvinism*. The Soviet system is deemed to be superior in all conceivable respects, including science and technology.

In the socialist countries the discussion of matters related to science and technology as a rule falls within the broad concept of 'scientific and technological revolution'. In the early 1960s the concept found its way into the Programme of the Communist Party while at the same time science itself was ideologically promoted to a higher rank by acquiring the status of a 'productive force', with the implication that it became an intrinsic part of the 'material basis' and hence ceased to be a mere aspect of the 'superstructure'.[2] But even before this ideological upgrading, science has always been regarded as something that forms an integral part of communism. Marxism itself allegedly represents an application of scientific method to political and economic practice (which is one of the most important reasons why its adherents consider it so vastly superior to the idealistic 'bourgeois' theories of social and economic structures). Thus, historically speaking, there has always existed an organic linkage between Soviet communism and science. The present-day leadership is officially thought to be presiding over the stage of the so-called 'advanced' or 'mature' socialism which alone can provide optimal conditions for full and rational utilisation of the fruits of scientific and technical knowledge.[3] By the same token, the theory goes on, capitalism is unable to provide these, and this implies that the Soviet leaders do their best to avoid a full evaluation of the extent to which the economy benefits from the importation of Western equipment because such an admission would objectively undermine the claims made by the ideological tenets of the socialist order.[4]

Roughly at the time when science and technology began to receive attention in the West as well as in the USSR the Soviet leaders, and notably Khrushchev himself, set out to travel widely throughout the Soviet Union, Comecon countries, Western Europe and even North America. In this way they acquired a first-hand opportunity to compare the levels of development among the nations and super-powers and thus to become aware of the technological gap that existed between the USSR and other socialist countries on the one hand, and the developed Western countries on the other. Khrushchev admitted that the USSR lagged behind the West but he did not lose his faith in the superiority of socialism as an economic and social system. Indeed, his faith was to some extent justified in view of the spectacular Soviet successes in the area of space exploration and in view of its strides in military technology. Those and similar achievements inspired Khrushchev's confidence that the technological gap could be bridged within a couple of decades. Small wonder that these extraordinary technological feats and claims created for some time a myth about Soviet technological might and that in the West there was some confusion as to the real strength of the USSR in the field of science and technology.

In spite of these claims and achievements, Soviet performance in science and technology may be judged to be disappointing. It has been established by Sutton that the Soviet Union depended on the importation of Western technology from its very inception. On the basis of extensive and detailed research into the role of Western technology in Soviet industrial development, Sutton comes to the conclusion that Soviet dependence on East-West technology transfer has been virtually total. He finds that between 1917 and 1965 there was no fundamental innovation of Soviet origin in the USSR and that recent evidence suggests that this unfavourable record is likely to remain unaltered.[5] A less sweeping conclusion, based on the 'most favourable' interpretation of the evidence amassed by the author, suggests that in the period 1917-30 no major technology originated in the USSR. In the period between 1930 and 1945 only two technological processes originated in the USSR, while in five areas the Russians developed and successfully applied some major technology which incorporated both Western and Soviet processes. In the period 1945-65 three processes were of Soviet origin and five technological areas used both Soviet and Western processes.[6] A 'less favourable' interpretation could, according to Sutton, accurately argue that those Soviet-developed processes were later replaced by ones of Western origin or, if used, they are relatively less efficient and their application is limited. A good example of this is the turbo-drill.[7]

Sutton, however, draws attention to two interesting facts that emerged from his research. Firstly, as far as the lack of data permits one to judge, it seems that the insufficient ability to innovate in the civilian sector has been to some extent offset by effective innovation in the defence sector. Secondly, the method by which the Russians do achieve some significant advances in civilian industry is basically that of scaling-up Western technologies. The indigenous technological progress (scaling-up) took place in three industrial areas: iron and steelmaking (but not steel rolling), electricity generation and high-voltage transmission, and rocket technology. But all these three industrial sectors as well as the defence sector may be singled out as traditionally high-priority areas and for this reason they cannot be regarded as being propelled forward by normal economic pressures. We have argued in the preceding chapter that this is precisely what the system is well capable of but it can do it only at the price of neglecting other areas. As regards the defence sector Sutton argues that since normal economic innovation has no clear-cut technical objectives it results from positive interaction of a myriad of complex forces and its success or failure can be tested only in the market-place (which the USSR does not possess). Military innovation, on the other hand, is predicated upon well-defined objectives. Military planners, unlike economic planners, can estimate fairly accurately what the next technological stage will be for a given weapon and can define a technical objective for that weapon in clear terms. In addition the system can provide all the necessary industrial back-up to the defence sector.[8] As far as the other three industrial sectors are concerned it is interesting to note that all three were at some time selected for prominent treatment by the leaders. Lenin's favourite was the GOELRO plan for the electrification of Russia. Stalin, as his name implies, favoured the iron and steel industry, and Khrushchev drew his faith in the strength of socialism from the success in rocketry and space exploration with which he was personally involved. The method of scaling-up of, for example, the size of open-hearth furnaces or of constructing longer and longer electricity grids for higher and higher voltages also goes hand in hand with the Soviet predilection to build things large and to pay little heed to the human scale. Marxist-Leninist economics, as opposed to classical economics, is not too concerned with factors conditioning the 'optimum size' of an undertaking. The centralised political power in a country with huge monotonous terrain becomes the victim of a tendency towards 'gigantomania' at the expense of optimality.[9]

The Soviet Union acquired Western technology by various means. During NEP and up to the late 1930s Western companies participated in Soviet industrial development through the policy of concessions. Plant and machines were imported together with Western engineers who supervised their erection and trained the Russian workers and engineers. Another channel of technology transfer was the acquisition of technical documentation or learning about new technology by 'reverse engineering', i.e. by dismantling a prototype purchased in the West. During World War II the Soviet Union received considerable technical assistance under the provisions of the Lend-Lease treaty with the Western Allies. After the war the Soviet Union dismantled a large number of industrial complexes situated in the occupied areas of Germany and Manchuria. In some cases this involved the 'acquisition' of scientists and researchers. The Soviet successes in rocketry, for example, were assisted by the USSR's ability to draw on the work of German scientists and on the relatively advanced research achieved in this sphere in Germany at the end of the war. During the period of the Cold War the Soviet Union was to some extent cut off from the technological advances accomplished in the West owing to the restrictions of the strategic embargo supervised by COCOM. This period, however, was relatively short and it has been argued that the impact of the embargo was not great. During the Korean War, at the height of the Cold War, the embargo applied to about 30-50 per cent of all items entering world trade and at the most it affected five per cent of Soviet defence and investment resources. The retardation effect produced during the decade of the 1950s was less than four months.[10] In the early 1960s, however, East-West trade began to grow and technological co-operation intensified with ever increasing liberalism applied on the Western side of the divide. Needless to say, at the same time the Soviet Union benefited from co-operation with the relatively developed East European Comecon countries, notably the GDR and Czechoslovakia, whose industries were geared to satisfy the Soviet market.

B The technological gap

Having considered the extent of Soviet dependence on acquisitions of Western technology in the course of its industrial development we can now turn to the problem of ascertaining the nature and size of the technological gap that separates the USSR from the advanced industrial countries of the West.

A word of caution may be in place at the outset of an assessment of the size of a technological gap. It is important to realise that any discussion of the gap (or lag) must inevitably to some extent suffer from a certain amount of ethnocentricity; that is to say, we tend to compare the technological level of the United States and the other Western countries with the level achieved by the Soviet Union and the other socialist countries without paying due attention to the fact that the latter may not necessarily be following the same path of development as the former. No two countries and their social and political systems are exactly alike, each pursues its own specific course of development and, consequently, any statement as to the size of technological gap existing between them must be understood to be applicable in only a very general sense. Indeed, it could be argued that the Soviet Union and the other Comecon countries have entirely different social, political and economic objectives from the United States and Western Europe and that this cannot be unreflected in their overall policy in the area of technological innovation and its application. The USSR, for example, is endowed with vastly larger geographical space and raw materials and fuel deposits than, say, Western Europe, and this factor certainly accounts for some of the profoundly different attitudes towards technology and its role in economic life.

Similar problems arise when we wish to compare the technological sophistication of two machines, one made in the Soviet Union and the other, for example, in the United States. Both machines may have broadly similar operating characteristics and more or less the same quality of their end products. But there may be a number of indices, such as reliability, rate of depreciation, energy consumption per unit of output and others, which may make any meaningful comparison between them hard to achieve. For instance a machine deemed to be producing a slightly better quality product than the other may be prone to more frequent technical breakdowns or may consume more energy. If reliability or energy savings were our objectives then we would be justified in choosing the machine producing the lesser quality end product as technically more successful and desirable than the machine producing a slightly better product at the cost of relatively higher energy consumption and frequent maintenance stoppages.

However, having cautioned against undue ethnocentricity it must at the same time be pointed out that we wish merely to warn that there cannot be any *absolute* yardstick for deciding which country is technologically more advanced than another, or which machine is technically more sophisticated than another. For practical purposes it is possible to assume on reasonably justified grounds that an industrially more developed country is technologically more advanced than a less developed nation. Moreover, if the declared aim of the socialist countries is to provide a better standard of living for their population than that achieved in the West, and if they claim that they can provide better conditions for full utilisation of scientific and technological progress, then we may be justified in assuming that some measure of universality in the paths of development must exist. Indeed, insofar as universality is considered to be one of the hallmarks of modern science it must be true that an inter-country comparison of performance in the sphere of science and technology is a meaningful exercise. How meaningful it is depends, ultimately, on the reader's point of view. In regard to the comparisons of particular machines or pieces of equipment, the experts and designers familiar with the technology in question are as a rule able to tell which of the two is technologically more advanced because they are in the best position to judge which indices of comparison are most important.

The simplest criterion for assessing the technological level of a country or of a given industry is labour productivity (i.e. output per man-hour of labour). This measure, however, is unsatisfactory because labour productivity increase may be due to a variety of factors, for example, to the extent to which capital is substituted for labour in response to changes in relative input prices, economies of scale and degree of capacity utilisation. A better method is the total productivity index. This method relates changes in output to changes in both labour and capital inputs. A still better method is the measure of combined factor productivity. It is based on the assumption that output is the function of the main factors of production – capital and labour. Any increase in output that cannot be explained in terms of increases in factor inputs can then be attributed to a *residual* combination of influences related to technological progress. Since the major impact of technological change is that it brings about a change in the production function,[11] economists are able by the use of a production function analysis to measure the variations in the level of technology of a given country (or industrial branch) over time – dynamic efficiency – or to compare inter-country technological levels at a given point – static efficiency.

A considerable amount of work has been done in the West on measuring the efficiency of the Soviet economy using different methods and approaches.[12] One of the best known studies is that of Boretsky, who compared the USSR with the United States on the basis of a number of technological indicators, the impact of which was assessed in savings in man-years. Boretsky found that the index of technical progress between 1939-40 and 1962 for the USSR was 63 per cent of that of the United States or, in other words, that the USSR was some 25 years behind.[13] In the latter years of this period the USSR was catching up and consequently closing the gap. Boretsky's study is mentioned here only as an example of a macro-economic study based on econometric computations and also in order to illustrate what the verdict of such a study might be. The methods using production function have generally tended to be more and more sophisticated, achieving a greater and greater degree of disaggregation.

It has been pointed out by Amann that in spite of improvements in the macro-economic methods of measuring the Soviet technological level these studies suffer from a number of technical difficulties, namely: (a) what factors to adjust for in measuring capital and labour inputs and how to price these inputs in the USSR and Western countries in order to be able to make reasonably sound comparisons; (b) the extent to which the phenomenon of technological progress is to be considered as exogenous or endogenous to an economic system and its operation; (c) why one production function should be preferable to another; (d) whether simple measures of output in value terms capture the full effect of technical progress (technological improvement may, for example, lead to a reduced pollution of the environment and improved quality of life-style of the population). There are, moreover, additional problems (specific to the USSR and other socialist countries) which make the task of interpreting the *residual* even more complicated. Chief of these is the uncertainty of the extent to which productivity increase may be due not to technical change but to increased effort and motivation resulting from Party campaigns, socialist competition or simply from reorganisation of enterprises into production associations.[14] Amann concludes that while considerable variations, depending on the type of the production function used and method of a particular study, exist, the general pattern that emerges is that the *rate* of technological development in the post-war Soviet Union has been superior to that of the United States and roughly on a par with other Western industrial countries, while the *cumulative* level of technology continues to be inferior to all major Western countries particularly to the US.[15]

An entirely different approach to that based on the use of the production function has been adopted in a major recently-published study of the technological level of Soviet industry collectively written by a team at the University of Birmingham.[16] The authors selected nine industries considered to be of key importance and undertook case studies of each of them comparing their technological levels with those of the relevant Western industries. This approach uses a number of suitable criteria by which to assess and compare the technological levels of the industries studied. These criteria include level of research (number of papers, citations, Nobel Prizes awarded to Soviet scientists), inventiveness (patents registered at home and abroad), innovation (comparing dates of first commercial production of a given product in different countries, comparative parameters of key items of equipment, technology transfer as a substitute for domestic innovation), and diffusion (comparative rates of diffusion of key technologies, pattern of foreign trade in industrial output). In this way, to use the expression of one of the authors, we are able to obtain a 'cluster of insights' into the technological structure of Soviet industry or, alternatively, we are able to observe 'manifestations'[17] of the technological level of Soviet industry and consequently to compare it with the West. The industries subjected to this kind of scrutiny include: high voltage power transmission, the chemical industry, computers, the iron and steel industry, machine tools and numerically controlled machine tools, passenger cars, military technology (tanks and inter-continental ballistic missiles), space rockets, and industrial processes control instruments.

The conclusions of the study are that the high priority sectors perform relatively well in the innovation phase. For example, in terms of thrust power Soviet rockets were more advanced than American until 1967 and the Soviet tank T 34 was for some time regarded as in many respects the best in the world. In civilian technology the study confirms the conclusions reached by Sutton in that the relative capability of the Soviet Union to develop new technology and bring it to commercial application appears most pronounced in the less R and D intensive industries with traditional ideological backings: the steel industry and electricity generation and transmission. The industries in which the Soviet Union encounters the greatest difficulties have tended to be the more R and D intensive ones such as computers, the chemical industry and numerically controlled machine tools. The development of these industries was for a considerable length of time neglected. The construction of a modern petrochemical industry, for example, did not take off until 1958 when Khrushchev launched his 'chemicalisation' drive and in this, as well as in the other research intensive industries, the closing of the gap necessitated a heavy dependence on technology imports. The overall conclusion of this study is that 'in most industries studied there is no evidence of a substantial diminution of the technological gap between the USSR and the West in the past 15-20 years either at the prototype/commercial application stages or in the diffusion of advanced technology'.[18] The general impression of the research team was that in considering the constituent phases of the 'research-products cycle' for each of the case studies conducted (i.e. research – experimental development – innovation – diffusion) the Soviet Union performs relatively well in the early stages. The volume of scientific work is substantial by any standards and is exceeded only by the United States. The quality of scientific output, however, is worse than this considerable effort would suggest. Other studies seem to confirm the overall conclusion that the Soviet Union is very weak in the latter stages of the cycle. In his study of the diffusion of imported Western technology within the Soviet chemical fertilisers industry Hanson has shown that the Russians are not as good at diffusing this technology as has generally been thought. The sluggish performance in diffusion of imported technology is probably attributable to the same causes that prevent the diffusion of technological innovation generally, regardless of whether it is of foreign or domestic origin.[19]

An important question now arises of whether the findings of the macro-economic studies based on the use of the production function (namely, that the rate of technological progress in the USSR has been superior to that of the US and Western Europe) can be reconciled with the conclusion reached by the Birmingham team (i.e. that there are no signs that the gap is narrowing). This question is especially pertinent in view of the opinion advocated in recent years by Gomulka, who not only maintains that the technological gap between the East and West has been narrowing but, moreover, foresees that this process will continue in the future. Apart from the fact that Gomulka is using a macro-economic approach, as opposed to the disaggregated case study method chosen by the Birmingham study, he introduces another important element into the picture – the West to East flow of technology. The West is thus seen primarily as the donor of technology with the USSR (and the other socialist countries) as recipient whose technological progress depends to a considerable degree on imports of technology from the West. According to this view, the Soviet output growth rates will foreseeably decline, as they have done in the past, to the levels recorded by the industrialised Western countries while labour productivity growth will settle to a low but steady rate of increase. But this process of retardation of growth is caused by the narrowing of the technological gap itself. In other words, by engaging in the imporation of Western technology the Soviet Union and the other socialist countries succeed in narrowing the gap, but by the same token they become more like the Western countries in the pattern of combination of their factors of production and in their mode of economic growth. However, an important point to be stressed here is that this model implies that in spite of the gradual narrowing of the technological gap it will never be entirely bridged because the Soviet Union will continue to occupy the position of a recipient and the West will remain the technological donor. (In part, the USSR, by paying for Western technology, contributes to further R and D efforts which, however, take place outside its borders, thereby perpetuating its position of dependency.) Ultimately the size of the gap will be determined by the extent to which the Soviet Union manages to open its borders to foreign investment and technology inflow and to open its system in both socio-political and economic terms, in other words, by the extent to which it manages to implement reforms of the existing arrangement.[20]

The implications of Gomulka's model may be, but are not necessarily, in conflict with the findings of the study written by the University of Birmingham group, since the former deduces a reduced technology gap from a reduced GNP per head of labour force gap, but aggregate labour (or total factor) productivity in the Soviet Union could come closer to the Western level while productivity levels in individual industries did not come any closer if the branch-composition of output and the branch-allocation of factor inputs were shifting faster in favour of higher productivity branches in the USSR than in the West. Be that as it may, Gomulka's approach clearly highlights the importance of the international flow of technology in considering the outcome of the long-term development of the Soviet and East European economies. (We shall return to this theme in the concluding part of this study.)

The remainder of this chapter is devoted to an analysis of the reasons why the Soviet Union's performance in the generation and diffusion throughout the economy of technological change has been so disappointing in spite of considerable human effort and investment. First we shall analyse the shortcomings attributable to the manner in which the R and D sector is organised, and then we shall consider the obstacles to technological change and its diffusion produced by the economic system itself.

C Organisation of Science and R and D in the Soviet Union

The overall policy is formulated by the Politbureau and is subsequently implemented by the central planning bodies, industrial ministries, the Academy of Sciences, the State Committee for Science and Technology (GKNT), and universities and other institutes of higher learning. The Academy of Sciences with its vast network of research institutes is the oldest organisation concerned with scientific research in the Soviet Union. It was founded in 1725 by Peter the Great in order to promote the application of science in industry and thus to satisfy the growing demands of the Russian army. However, throughout its history the Academy has tended to focus mainly on fundamental research which may perhaps be seen as both the cause and the consequence of the traditional Russian neglect of applied industrial scientific activity and industrial backwardness in general. During the Soviet period the Academy was moved from Leningrad to Moscow and subjected to several reorganisations. It was also substantially enlarged and various attempts were made to make it more oriented to applied research. This culminated during the war when many institutes belonging to the Academy were directly involved in an intensive programme of military research. Today the Academy remains the highest scientific establishment in the Soviet Union. Its basic objectives are fundamental research in natural and social sciences. It is under the control of the Council of Ministers but it has general control over investigations of the most important scientific problems throughout the USSR by virtue of its supervision of the work of its republican branches. The Academy has a network of research institutes, laboratories, experimental stations, libraries, etc., all of which are organised within a structure not unlike that of the industrial ministries: their work is subjected to the process of central planning and the individual research institutes may be likened to industrial enterprises (who represent the lowest rung of the ministerial system).

The second major structure of organisation of scientific research and development is the institutes belonging to the industrial ministries themselves. Initially the system of industrially oriented institutes worked and gradually expanded under the direction of the Scientific Department of the Supreme Economic Council of the National Economy. In the early 1930s under the impact of rapid industrialisation this system of research institutes, design bureaux and capital-project institutes was incorporated into the industrial ministerial structure with each ministry taking responsibility for those agencies which served it. Their work is planned in harmony with the plans of their ministries by the Technical Administration of the relevant ministry. The Technical Administration, however, prepares its plan in response to the proposals of the Production Administration of the relevant ministry, and thus the planning of research and development is accomplished in a highly centralised manner, on a higher level, with the result that research activity does not directly respond to the immediate needs of the enterprises whom the research institutes are supposed to serve. The enterprises themselves often have their own R and D facilities, laboratories or experimental shops, but traditionally these facilities have suffered from the lowest priority in terms of supplies and overall importance. The system that has evolved in the USSR favours the notion that scientific and technical research should be undertaken in isolation from the process of production both in organisational and geographical terms (institutes are often situated in large cities far away from the factories whose problems they are helping to solve). These structural and organisational faults have been recognised as one of the main causes of the unsatisfactory rate of diffusion of innovation in Soviet industry, and the recent reforms of aggregating enterprises into larger units, the production or industrial associations, involve the transfer of some of the research institutes to be under the direct supervision of the associations with the aim of overcoming the organisational barrier.[21]

Soviet universities and institutes of higher learning play a very limited role in relation to the overall scientific effort of the USSR. This is in sharp contrast with the situation existing in the West, where universities as a rule fulfil the function of research centres in addition to their teaching role. Western commentators often interpret this tendency as a consequence of the desire to prevent mushrooming of independent research bases (especially in social sciences) and thus of potential sources of dissent. However, some amount of interlocking with the industrial sector does take place and leading researchers often also teach.[22]

The main features of the organisational structure of Soviet scientific and technical research may, on the basis of the above outline, be described as the vertical separation of research and development from production and the horizontal division of research and development activity between the Academy system, industrial sector and higher education system.[23] In addition there exists further fragmentation within the industrial sector as a consequence of the departmentalism and autarkic tendencies characteristic of the Soviet system. Although there have been various reorganisations and changes in policy from time to time, the system that evolved during the industrialisation has changed only marginally and the fragmentation of scientific activity persists.

One of the most important events in the evolution of the Soviet science system was the creation in 1965 of the GKNT. It was founded as a successor to the earlier organisation, the State Committee for Co-ordination of Scientific Research, and its main purpose has been to help overcome the fragmentation of research efforts which had become increasingly intolerable as the crash industrial programmes, such as 'chemicalisation', became more and more research intensive and research itself more expensive.[24] The task of the GKNT is to ensure the cohesion of state policy in the field of scientific and technical progress and to determine the basic tendencies in the development of science. In addition it organises the dissemination of scientific information and has an overall responsibility for increase of efficiency of scientific activity and for speedier introduction of new technology in industry. The GKNT participates together with *Gosplan, Gossnab*, the Academy of Sciences, ministries and other organisations in the process of allocation of resources for individual projects and in supervising the implementation of governmental scientific policy.[25] The Committee has a number of institutes, the main one of which is the All Union Institute of Scientific and Technical Information. Other important functions of the GKNT include the popularisation of scientific knowledge, organising of exhibitions and provision of information about what is going on abroad. It also plays a major part in the process of acquisition of Western technology by virtue of its monitoring of Western progress and by making recommendations as to what ought to be purchased. One of the key figures in this respect is Z. Gvishiani, a vice-chairman of the General Assembly of the GKNT, who is generally regarded as the most eminent Soviet authority on American studies in management and one of the most active proponents of the adaptation in the USSR of Western approaches to industrial management.[26] The creation of the GKNT resulted in a certain amount of friction within the organisational hierarchy of Soviet science. The Academy feels threatened by this younger and often very influential body, and in any case the creation of the GKNT terminated the exclusive right of the Academy to make all science decisions and to represent Soviet science abroad. Many industrial ministries too have had clashes with the Committee since the latter often insists on a speedier introduction of new technology within an industry, much to the discomfort of the bureaucracies involved.

D The main causes of the disappointing performance

The Soviet Union embarked upon the road of abolishing market relations and with it all the uncertainty, price fluctuations, strikes, unemployment, cyclical crisis, and anything else we wish to name. The centrally planned system which replaced this 'anarchy' gives a strong support to stability, but innovative change by definition involves disruption of the existing *status quo* and is, therefore, in conflict with the objectives and needs of the system. Spontaneity and economic autonomy of the small and large units operating within the competitive framework of a market economy are powerful factors stimulating constant search for innovation and ensuring its speedy introduction. To the extent to which the Soviet Union and its socialist allies succeeded in conquering the forces of free market and substituting for its mechanism stable, centrally planned direction of development, they must pay their price in terms of losses attributable to an insufficient rate of technological progress. Moreover, the desideratum of high growth-rates of output of a relatively limited group of commodities in the absence of consumer sovereignty further fortifies technological conservatism, as there is, so to speak, no time to be spared for the act of introducing an innovation. It is almost as if the considerable macro-innovation, accomplished by embarking upon the road of construction of a new social and economic order, leaves no room for promoting innovation on the micro-scale. All these factors, then, have left the leaders and planners without effective tools which they might otherwise have possessed. The instruments peculiar to the centrally directed system – an increased ability to introduce innovation from above by administrative means or by continuous exhortation – have so far unfortunately failed to demonstrate in a convincing manner their much hoped for superiority over the spontaneity of the market. The centre simply cannot oversee everything at the same time and, consequently, it cannot enforce its wishes along the whole front of economic activity in an optimal manner.

In the most comprehensive study to date, Joseph Berliner[27] provides an excellent exposition of the intricacies of the Soviet system that impede introduction of innovation in industry. The author identifies at least four independent structures which in one way or another affect the decision to innovate at the level of enterprise (he treats the enterprise as the centre of his study) – the organisational structure, the prices structure, and the structures of rules and incentives. Berliner concludes that the system is designed in such a way that introduction of innovation brings about a number of risks to the innovating manager and that the incentives or likely benefits obtainable from successful introduction of innovation are not sufficiently substantial to motivate the persons responsible for taking the decision. The Soviet system lacks a more favourable, pro-innovation, balance of risks and rewards and thus it finds it hard significantly to improve its performance. Adam Smith explained the importance of the 'invisible hand' in the process of creating wealth but the advent of the scientific and technological age clearly focussed attention on the fact that the modern economy, dependent on intensive factors of growth, requires in addition what Berliner aptly describes as the 'invisible foot' or, in other words, a mechanism which does not simply sufficiently reward an innovator, but which at the same time punishes the manager who does not innovate. The Soviet Union has so far failed to introduce and successfully operate such a mechanism although the topic has been discussed at least since Khrushchev's time and some marginal reforms have been implemented. However, much more will have to be done if the USSR is to free itself from its technological dependence on advanced Western countries. Berliner believes that contrary to the view frequently expressed by the Russians themselves, there is nothing inherent in the character of the Soviet nations that makes the promotion of innovation and greater industrial efficiency less possible than elsewhere in the world. The causes of insufficient efficiency in the area of innovation are entirely systemic and some reforms of the system could be achieved even without offending the ideological precepts on which the system rests.

The situation as it is outlined by Berliner[28] shows a centrally planned system stipulating what the enterprise should produce and telling it where to obtain the inputs, but we have seen that owing to inevitable imperfections in these horizontal flows of resources the enterprises tend to resort to autarky and hoarding to cushion themselves against the harshness of their environment. Under these conditions the managers learn to value, above all, stability and well-established sources of supplies. Unfortunately the decision to innovate is often likely to increase the intensity and complexity of the supply system of an enterprise. It requires the use of new materials with which the enterprise is not familiar, to be supplied by new suppliers whom the management does not know and who in their turn do not know the quality, and other requirements of the receiving enterprise. Introduction of a new technology or of a new product always involves a certain amount of wastage and disruptions until the new process is mastered, and if this is not foreseen in the plan the manager is likely to suffer in respect to superior authorities. A related reason why the task of introducing an innovation under the conditions of the Soviet system is so complicated is the fact that the enterprise requires the services of an R and D organisation. We have seen that the system maintains an organisational barrier between the R and D sector and the process of production, and thus the risk of failure implied in dealing with a research institute is substantially more acute than if there existed more direct horizontal linkages. The problem of supplies both in the material sense and in terms of R and D services does, of course, also arise under the conditions of a market economy. Within the operational paradigm of a Soviet-type economy, however, the problem turns into a major one because of the monopolistic position of the suppliers in what basically is a sellers' market. The manager has nowhere else to turn in the event of shoddy deliveries or services on the part of his suppliers and R and D agencies. Recently implemented reforms include the above-mentioned merger of enterprises into production associations. The most important development in this respect is the creation of so-called 'science-production associations', a type of merger involving an aggregation of a large research institute with several enterprises with the result that the head of the institute assumes the directorship of the whole association. Another important recent reform is an attempt to introduce greater flexibility into the wholesale supply system by rescinding the old rule requiring all supplies to be allocated in compliance with centrally issued material allocation certificates. So far, however, the implementation of such reforms has not progressed with sufficient speed. So much for the organisational structure of the Soviet economy and the way in which it influences the decision to innovate.

The disincentive to innovate generated by the organisational structure could to some extent be compensated for if the prices of new products were allowed to rise in proportion to the risks the innovating manager has taken by virtue of his decision to introduce new technology. Berliner argues that the principle of determining prices, under the influence of one of the tenets of Marxism – the labour theory of value – in practice brings about a situation where all products' prices are set in the same manner, regardless of whether the product is new or old. The profit mark-up is determined by the average cost of production, which means that a new product embodying a technological advance not possessed by an older one will be assigned the same price as the older one. This discriminates against the producer of new equipment for he has nothing to show for the risks he has taken. Some reforms of the price system with a view to ending this type of discrimination have been implemented, but the notion of pricing products in accordance with their productivity goes very much against the grain of the ideology and is therefore resisted.

The Soviet system's structures of incentives and rules governing the behaviour of an industrial manager are closely interrelated by virtue of the fact that his following of certain rules, deemed necessary or desirable by the central authorities, is encouraged by a corresponding system of financial incentives.[29] A brief discussion of the structure of incentives should, therefore, suffice for our purposes here. During the industrialisation of the 1930s the leaders introduced a system of incentives designed to induce managers to maximise output, and there were no incentives for a successful introduction of innovation. It was not until the mid-1950s that for the first time a special bonus was introduced with the specific purpose of stimulating industrial innovation. The size of this bonus is regulated by the reduction in production costs that the innovation brings about. Further measures of a similar kind have been introduced since then so that the system of incentives is more and more geared to meet the requirements of the intensive strategy of growth. However, Berliner has shown that this movement has not gone far enough to be really effective. The bonus earned for fulfilling the plan continues to be the major source of financial incentives of the managers, and since innovation continues to have a negative effect on the enterprise plan fulfilment the innovating enterprise earns somewhat smaller bonuses than an equally well managed non-innovating enterprise. The innovation bonuses are of such magnitude that as a rule they merely compensate for the loss of some part of the basic bonus. The balance of rewards for risks continues to be tilted in innovation's disfavour.

* * *

The causes of technological conservatism exposed by Berliner pertain merely to one aspect of the system and thus, however important the enterprise may be as the basic unit of the economy, the analysis concerns only one facet of the whole problem. Whichever way we look we are bound to identify a whole range of other causes.

To start with, the very unpredictability of scientific discovery seems to rule out the notion that scientific knowledge can best grow under the conditions produced by the centrally planned system. Although planning is an ideal tool for the purpose of allocation of resources to individual research projects, especially in view of the fact that scientific and technical research has been becoming more and more resource-demanding, the actual rate of 'scientific output' cannot be subjected to planning in the conventional sense applicable to the rest of the economy. Those who argue that strict central planning as a method of steering an economy is in practice less efficient and less rational than reliance on market-based relations probably have twice as strong a case when their arguments are applied to the question of whether central planning represents an optimal method in the direction of scientific research and its speedy application in the process of production. Scientific activity is even more unpredictable than the outcomes of conventional, everyday decisions, taken in normal economic life.

A major systemic feature inhibiting the attainment of a satisfactory rate of technological change is the considerable bureaucratisation of life. Under the conditions of central planning nobody wants to take a decision without having confirmed its propriety with his superior organs and, consequently, every new measure has to come up and down the endless bureaucratic channels. But the success of an innovation frequently depends on the degree of freedom to respond quickly to the specific requirements of a given problem situation. Any bureaucracy must continually justify its existence by 'supervising' the activity of the lower bodies in the command system, and this inevitably creates a situation where the higher authorities, unaware (and often not bothering to be aware) of the specific situation below, frequently interfere insensitively in the decision-making process of the lower bodies, thereby not only curtailing the personal initiative of the more enterprising industrial managers, but causing more problems rather than helping to solve them.

Another drawback of the Soviet system is that in spite of the substantial changes accomplished since Stalin's death the USSR is still by any standard a very closed country, curtailing freedom of discussion, access to information and the right to free travel both within the USSR and abroad. But inventiveness and diffusion of technological change vitally depend on these freedoms. If the experience of the majority of industrially highly advanced nations is anything to go by, it would suggest strongly that scientific and technological progress and basic civil liberties may in the long run be judged to be inseparable. Undoubtedly the granting of these freedoms constitutes to some extent a double-edged sword. The United States, for example, has benefited from the 'brain drain' to the detriment of other countries who financed the training of the scientists migrating there. The Soviet weapon against the brain-drain is the exit visa, but by closing its frontiers the USSR cannot attract brains into what many regard as a highly oppressive system.[30] The same applies to the other socialist countries of Comecon: the emigration of, above all, professional and qualified manpower from the German Democratic Republic to the West was brutally halted in August 1961 by the erection of the infamous Berlin Wall. Technical progress in the GDR subsequently improved, but such a drastic solution surely cannot be regarded as adequate in the long run. Instead of increasing their power of attraction the socialist countries have chosen to rely on oppression. This renders them still less attractive, and their chances of benefiting from the process of international dissemination of knowledge become ever more distant.

Lastly, we might mention the negative influence of ideology in the realm of science. The existence of the official Marxist ideology may be regarded as the main cause of the above-mentioned absence of free discussion and access to information. Virtually all branches of Soviet science have suffered from the consequences of the imposition of a single *Weltanschauung* in all areas of scientific enquiry. Even research in pure and applied sciences (and doubly so in the humanities) may raise questions, or produce implications, that ultimately clash with the precepts of the official doctrine. The disciplines that have suffered most are the social sciences, economics, linguistics and others. Cybernetics, for example, was for a considerable length of time regarded as a bourgeois pseudo-science, and this may largely account for the relatively slow take-off of computerisation and automation in the USSR. Ideological strictures may also to some extent be held responsible for such causes of technological conservatism as, for instance, the slow rate of dissolution of obsolete factories. The high rate of technical change achieved in the West has to be paid for in terms of unemployment. This, of course, would be considered unacceptable by the socialist countries. Moreover, for purely practical reasons such unemployment would be harder to cope with and therefore ideologically and socially more offensive: temporary non-employment caused by the scrapping of productive capacity gives rise to considerable problems even under conditions of relative scarcity of labour because the socialist countries do not possess the necessary superstructure making the movement of manpower sufficiently flexible.

On the other hand, ideology may be instrumental in promoting technological progress in some sectors (as has been the case in the branches traditionally supported by the leadership which we have discussed above), but the benefit derived in this way is strictly limited because such selective promotion is undertaken at the cost of neglecting the rest of the economy. This is yet another manifestation of the system's lop-sidedness discussed in the preceding chapter. The strength achieved in one direction leads to a weakness in another but, unfortunately, any measure of spill-over of knowledge or qualified personnel from the relatively advanced sector – the defence-related industrial complex – is by and large prevented by the prevailing atmosphere of secrecy.[31] There is a barrier between the civilian and defence sectors in the USSR which in the West is often described as the 'internal CoCom'. Similar barriers also exist within and between the individual branches of the civilian sector of the economy and are entirely due to the autarkic tendencies, organisational fragmentation and departmentalism deeply ingrained in economic and political systems of the Soviet type.

Notes

1 P. Wiles, 'World Technological Leadership', *Lloyds Bank Review*, no.127, January 1978, p.25.

2 L. Labedz, 'Science and Soviet System', ibid., p.149.
3 A.B. Evans, 'Developed Socialism in Soviet Ideology', *Soviet Studies*, no.3, 1977, pp 409-28.
4 Z. Nagorski, *The Psychology of East-West Trade*, New York, 1974, p.205. The author points out that the impression a visitor gets at Togliatti is that the Fiat car plant constructed there is entirely a Soviet undertaking. In socialist countries the subject of technological co-operation with the West is subjected to censorship.
5 A.C. Sutton, *Western Technology and Soviet Economic Development*, Stanford, 1973, vol.III, introduction, p.xxv.
6 Sutton, ibid., vol.III, p.370.
7 Ibid.
8 Sutton, ibid., pp 361-2.
9 R. Hutchings, *Soviet Science, Technology, Design: Interaction and Convergence*, OUP, 1976, pp 215-16.
10 G. Adler-Karlsson, *Western Economic Warfare 1947-1967*, Almqist and Wiksell, Stockholm, 1968, pp 15 and 191.
11 E. Mansfield, *The Economics of Technological Change*, W.W. Norton, New York, 1968, p.14.
12 Useful surveys of this literature appeared in Amann, Cooper and Davies, *The Technological Level of Soviet Industry*, Yale UP, 1977, chapter 1 and M. Kaser, 'Government and Innovation in the Soviet Union', *Annuaire de l'URSS et des pays socialistes européens*, Centre de Recherches sur l'URSS, Université des Sciences Juridiques, Politiques et Sociales de Strasbourg Institut de Recherches Juridiques, Politiques et Sociales, 1975, pp 32-6.
13 M. Boretsky, 'Progress in Technology, Productivity and Economic Efficiency: USSR versus USA' in *New Directions in the Soviet Economy*, JEC, 1966.
14 R. Amann, 'Soviet Technological Performance' in Thomas and Kruse-Vaucienne (eds), op.cit. (chapter 1, note 19), p.331.
15 Ibid., p.332.
16 Amann, Cooper and Davies, op.cit. (above, note 12).
17 Amann, loc.cit. (above, note 14), p.332.
18 Amann, Cooper and Davies, op.cit. (above, note 12), p.66.
19 P. Hanson, 'The Impact of Western Technology: A Case Study of the Soviet Mineral Fertiliser Industry', paper presented at the Conference in Integration in Eastern Europe and East-West Trade, Bloomington, Indiana, October 1976.
20 S. Gomulka's work includes the following papers: 'Inventive Activity, Diffusion and the Stages of Economic Growth', Aarhus, 1971; 'Import of Capital Goods, Technical Change and the Rate of Development', a paper presented to the Oslo Econometric Meeting, August 1973; 'Soviet Post-War Industrial Growth, Capital-Labour Substitution

and Technical Changes: a Re-examination' in Z. Fallenbuchl (ed.) *Economic Development in the Soviet Union and Eastern Europe*, Praeger Publishers, 1976, vol.2, pp 3-47, and Gomulka and J.D. Sylvestrowicz, 'Inter-Country Embodied Diffusion and Time Changes in the Factor Productivity Residual', a paper presented to the Reisenburg Symposium, 1974.

21 A. Gorlin, 'Industrial Reorganisation: the Associations' in *Soviet Economy in a New Perspective*, JEC, 1976, p.166.

22 J. Turkevich, 'Soviet Science Policy Formulation' in Thomas and Kruse-Vaucienne (eds), op.cit. (chapter 1, note 19), p.43.

23 R. Amann, 'The Soviet Research and Development System: the Pressures of Academic Tradition and Rapid Industrialisation', *Minerva*, Macmillan (Journals), vol.VIII, no.2, April 1970, p.220.

24 Ibid., p.221.

25 J. Turkevich, loc.cit. (above, note 22), pp 24-5.

26 Yakov M. Rabkin, 'Science Studies as an Area of Scientific Exchange' in Thomas and Kruse-Vaucienne (eds), op.cit. (chapter 1, note 19), p.75.

27 J. Berliner, *The Innovation Decision in Soviet Industry*, MIT Press, 1976.

28 A concise exposition of Berliner's findings appeared under the title 'Prospects for Technological Progress' in *Soviet Economy in a New Perspective*, JEC, 1976, pp 431-46.

29 Berliner, *The Innovation Decision in Soviet Industry*, p.402.

30 P. Wiles, 'World Technological Leadership', loc.cit. (above, note 1), p.23.

31 On Soviet obsession with secrecy see Zhores Medvedev, *Soviet Science*, OUP, 1979, pp 121-6. See also his comments on what in this study we refer to as 'lop-sidedness', ibid., p.139.

PART II

3 Soviet chemical industry and East-West transfer of technology

One of the most striking features of the post-Stalin strategy of industrial development in the USSR (and in the rest of Comecon) has been a rapid expansion and modernisation of the chemical industry. It is in keeping with the nature of the Soviet system that the programme of modernising the industry took the form of a major industrial campaign, the so-called 'chemicalisation' drive. The campaign was launched by Khrushchev in 1958 and today it remains one of the few of his major policies not subsequently discarded by his successors: the expansion of the chemical industry has continued unabated since 1958 and there are no indications that the priority the industry has enjoyed is going to be diminished in the foreseeable future. 'Chemicalisation' was facilitated by Khrushchev's pursuit of the policy of 'peaceful co-existence' with the capitalist West and by tacit recognition on the part of the Soviet planners that any such programme of radical modernisation cannot be implemented without much freer access to the fruits of technological progress accomplished in the West in the preceding decades. Today, after twenty years of 'chemicalisation', the technological ties between the USSR and Western companies are stronger than ever before and the original concept of peaceful coexistence has in the meantime grown into a more comprehensive and complex process of East-West *détente*.

The aim of the following two chapters is to illustrate the extent to which imports of Western technology have been instrumental in securing the success of 'chemicalisation'. The discussion covers the period starting from the inception of the campaign but the main emphasis is laid on the flow of technology which materialised in the 1970s largely because of the greater topicality of the issues arising out of the recently expanded East-West industrial co-operation. Above all, these include the consequences of the 'compensation', 'buy-back' or 'countertrade' agreements, a practice whereby Western suppliers of technology receive payments from socialist partners partially, or wholly, in products turned out by the imported equipment. 'Compensation' trading is discussed in chapter 5.

The method chosen in order to satisfy our aims is to present an outline of the structure of the Soviet chemical industry and to discuss some of the most important quantitative and qualitative changes it has registered during the period under review. With this information constituting the necessary background, an attempt is made to illustrate the way in which the Soviet chemical industry has benefited from imports of Western technology by showing how the imports fit into the developmental strategy pursued by the planners. Unfortunately, a precise evaluation of the benefits the Soviet Union has reaped through its policy of importing Western plants would hardly be possible within the confines of this study and, moreover, the information concerning Soviet purchases of Western chemical technology assembled here can by no means be regarded as completely exhaustive. Nevertheless, we hope that a meaningful picture illustrating the role of Western technology in the growth of the Soviet chemical industry will emerge and that a reasonably firm judgement of the nature of Soviet dependence on the West may be formed.

A Khrushchev's 'chemicalisation' and its meaning

In 1955 the Soviet Union announced its Sixth Five Year Plan which was intended to cover the period 1955-60. After several revisions of its over-ambitious targets, the whole plan had to be scrapped and the planning bodies were ordered to prepare the launching of a new plan, the Seven Year Plan, for the period 1958-65. In the meantime the economy was run on the basis of provisional yearly plans. The Seven Year Plan was finally adopted in January 1959 by the 21st Party Congress, an occasion which also marked the culmination of Khrushchev's rise to power. The plan called for a far-reaching programme of overall modernisation of the economy, and one of the most important facets of this modernisation consisted of a programme of wholesale transformation and expansion of the chemical industry.

'Chemicalisation' implied a rapid expansion of output of a whole range of synthetic materials previously not manufactured in the USSR such as polyamides (*kapron*), polythene, polypropylene, polyformaldehyde, polyester (*lavsan*), polyacrylonitrile (*nitron*), polystyrene, phenolpolyuretane and others.[1] Plastics, resins and synthetic fibres were to replace wherever possible the traditional materials such as wood, ferrous and non-ferrous metals, cotton and furs. In agriculture, chemical fertilisers were intended to play a major role in efforts to secure higher yields of grain and, ultimately, of meat and dairy products. The plan foresaw a quadrupling of the output of artificial fibres and a sevenfold increase in the production of plastics and synthetic resins. Table 2 presents the main targets of the Seven Year Plan for the Soviet industry as a whole and for the chemical industry separately.

Table 2

	1958	1965 plan	1965 actual
National income	100	162-165	158
Gross industrial output (index)	100	180	184
Producers' goods	100	185-188	196
Consumers' goods	100	162-165	160
Iron ore (million tons)	88.8	150-160	153.4
Pig iron (million tons)	39.6	65-70	66.2
Steel (million tons)	54.9	86-91	91.0
Coal (million tons)	493	600-612	578
Oil (million tons)	113	230-240	242.9
Gas (milliard cubic metres)	29.9	150	129.3
Electricity (milliard Kwhs)	235	500-520	507
Mineral fertilizer (million tons)	12	35	31.6
Synthetic fibres (thousand tons)	166	666	407
Machine tools (thousands)	138	190-200	185
Tractors (thousands)	220	–	355
Commercial timber (million cubic metres)	251	275-280	273
Cement (million tons)	33.3	75-81	72.4
Cotton fabrics (million square metres)	5.79	7.7-8.0	7.08
Wool fabrics (million square metres)	303	1,485	365
Leather footwear (million pairs)	356.4	515	486
Grain harvest (million tons)	134.7	164-180	121.1
Meat (total) (million tons)	3.37	6.13	5.25
Workers and employees (millions)	56.0	66.5	76.9
Housing (million square metres)	71.2	650-660*	79.2

	Planned output 1965
	000 tons
Ammonia	1,500/1,600
Caustic soda	1,300
Calcinated soda	3,500
Sulphuric acid	10,000
Synthetic rubber	650
Motor tyres	30
Mineral fertilizers	35,000
Artificial and synthetic fibres	650

*Total for seven years 1959-65

Source: A. Nove, *An Economic History of the USSR*, p.353.

The creation of the synthetic materials industry first of all required the implementation of a major shift in the structure of feedstock in favour of oil and natural gas because the modern chemical industry and its development largely hinges on the build-up of a petrochemical industrial base. The early 1960s, therefore, saw the first steps in the construction of the Soviet and Comecon system of gas and oil pipelines in order to cater for the growing needs of the East European and Soviet petrochemical industries (in addition to their growing energy requirements). The USSR and the other members of Comecon started to implement this strategy with a lag of more than ten years behind the advanced capitalist countries. As early as 1948 about 40 per cent of organic chemicals in the United States were based on oil and natural gas. In the early 1960s about 60 per cent in the United Kingdom and 80 per cent of French organic chemicals were based on these feed-stocks.[2]

By 1963, after the Soviet Union encountered a number of economic and political difficulties on both domestic and external fronts, it became increasingly apparent that the goals of the Seven Year Plan were over-ambitious and the plan had to be scaled down. However, even the revised version remained underfulfilled in a number of sectors. The actual rate of fulfilment ranged from 80 to 100 per cent. The subsequent plan, the Eighth Five Year Plan, covering the period 1966-70, fared much worse: it was fulfilled only 79 per cent in chemical fertilisers, 46 per cent in synthetic fibres, 48 per cent in plastics and resins and 78.6 per cent in tyres. After this sobering experience the planners opted for more realistic targets in preparation for the Ninth Five Year Plan (1971-76) with the consequence that this time the industry registered far more satisfactory rates of plan-fulfilment. Except in plastics and resins, where the fulfilment was 80 per cent, all targets reached 100 per cent fulfilment. The current plan, the Tenth Five Year Plan (1976-80) to some extent reflects the optimism brought about by the success of the previous period.[3] However, it is doubtful whether this plan will be fulfilled bearing in mind the overall slowdown of Soviet economic growth. Already the output of fertilisers, for example, lags behind the targets for 1976 and 1977. Table 3 presents the output data of major chemical products between 1958 and 1977. What transpires is that in spite of the difficulties, inevitably brought about by the growth of a new industrial sector, and in spite of frequent underfulfilment of the planned targets, the growth of the Soviet chemical industry throughout the period has been by all standards quite remarkable. By 1973 the Soviet Union had become the world's largest producer of chemical fertilisers, and in absolute terms it now possesses the second largest chemical industry in the world (the US being the leader), although in terms of *per capita* output of chemicals it still has a long way to go before it catches up with the levels reached by the Western industrialised countries.

The 'chemicalisation' programme required a tremendous injection of capital. The chemical industry's capital stock rose about ten times between 1957 and 1974, and the per annum rate of growth of fixed investment in the same period was 28 per cent. The proportion of capital expenditure allocated to enlargement or reconstruction reached over 50 per cent.[4] Table 4 depicts the effect of this investment in terms of expansion of productive capacity due to reconstruction or construction of new chemical complexes.

Table 3
Production of major types of chemical output, 1958-1977
(Units in thousands of metric tons except when indicated)

	1958	1959	1960	1961	1962	1963	1964	1965	1966	1967	1968	1969	1970	1971	1972	1973	1974	1975	1976	1977
Synthetic resin and plastic masses	258	293	312	404	473	589	720	805	971	1,113	1,291	1,453	1,673	1,864	2,042	2,320	2,493	2,842	3,058	3,309
Mineral fertilizer total millions of tons of which:	12.4	12.9	13.9	15.3	17.3	19.9	25.6	31.3	35.9	40.0	43.5	45.9	55.4	61.4	66.1	72.3	80.4	90.2	92.2	96.8
Nitrious (in conversion to sulphate ammonia) (million tons)	4.1	4.4	4.9	5.7	6.9	8.6	10.2	13.2	15.5	18.3	20.4	22.0	26.4	29.5	31.9	35.3	38.3	41.6	42.0	44.5
Phosphorous (in conversion to 18.7 per cent P_20_1) (million tons)	4.7	4.7	4.9	5.0	5.2	5.9	7.5	8.6	9.5	10.0	10.3	11.1	13.4	14.8	15.7	17.3	20.9	28.8	25.8	27.8
Potassium (in conversion to 41.6 per cent K_20) (million tons)	2.4	2.5	2.6	2.8	3.2	3.4	4.6	5.7	6.3	6.9	7.5	7.8	9.8	11.6	13.1	14.2	15.8	19.0	20.0	20.1
Phosphorous meal (in conversion to 19 per cent P_20_1) (million tons)	1.2	1.3	1.5	1.8	1.9	2.0	3.2	3.8	4.4	4.8	5.2	5.1	5.7	5.4	5.3	5.4	5.4	5.5	4.4	4.3
Chemical means of plant protection and pest control	26.6	30.7	35.8	42.4	57.7	61.3	85.3	103.2	114.6	123.1	135.8	157.0	163.8	154.8	173.3	200.5	231.6	264.0	276.0	290.0
Sulphuric acid in monohydric	4,803	5,082	5,398	5,718	6,132	6,885	7,647	8,518	9,367	9,737	10,159	10,665	12,857	12,779	13,685	14,855	16,663	18,645	20,015	21,104
Calcium soda, 95 per cent	1,692	1,728	1,887	2,115	2,332	2,545	2,740	2,871	2,963	3,169	3,293	3,462	3,668	3,820	4,052	4,149	4,484	4,692	4,842	4,876
Caustic soda, 92 per cent	709	757	765	897	961	1,049	1,153	1,303	1,393	1,524	1,658	1,813	1,938	2,028	2,064	(2,020)	(2,174)	2,395	2,604	2,658
Automobile tyres, million units	14.4	15.5	17.2	19.0	20.8	22.6	24.4	26.4	27.7	29.6	31.8	32.6	34.6	36.2	38.8	42.3	47.1	51.5	54.5	57.4
Synthetic detergents	7.9	10.3	22.9	26.6	50.1	55.1	100.5	144.0	189.0	248.0	346.0	417.0	470.0	496.0	533.0	610.0	656.0	769.0	828.0	824.0
Artificial and synthetic fibre	166.0	179.5	211.2	250.4	277.3	308.4	361.1	407.3	458.0	511.0	554.0	584.0	623.0	676.4	746.1	830.0	887.0	955.0	1020.0	1088.0
Total of which:																				
Artificial fibre	153.3	165.5	196.2	226.8	243.4	265.8	304.3	329.8	362.0	395.0	424.0	441.0	456.4	473.3	507.5	543.0	569.0	590.0	612.0	629.0
From this:																				
Artificial silk	41.8	43.3	47.4	49.7	54.5	63.2	75.5	78.6	82.0	87.0	89.0	91.0	93.1	98.2	105.4	110.0	111.0	122.0	126.0	–
Artificial silk for cord	37.9	44.6	53.7	55.2	58.4	68.2	76.7	86.0	99.0	104.0	117.0	127.0	129.0	132.3	140.0	149.0	156.0	159.0	160.0	–
Artificial staple fibre	73.6	77.6	95.1	121.9	130.5	134.4	152.0	165.2	181.0	203.0	218.0	223.0	234.3	243.0	262.0	284.0	302.0	309.0	326.0	–
Synthetic fibre	12.7	14.0	15.0	23.6	33.9	42.6	56.9	77.5	96.0	116.0	130.0	143.0	166.6	203.0	239.0	287.0	318.0	365.0	408.0	459.0
From this:																				
Synthetic silk	6.2	7.0	6.6	8.7	11.3	12.4	15.9	20.3	22.0	28.0	31.0	38.4	39.5	42.0	47.9	52.0	54.0	60.0	64.0	–
Synthetic silk for cord	(2.0)	2.0	2.1	3.7	5.7	7.6	12.0	(17.4)												
Synthetic silk for industrial articles	(0.9)	0.8	2.1	2.9	5.6	7.7	11.3	(16.5)	47.4	57.5	63.0	67.0	75.6	90.3	109.3	124.0	141.0	171.0	196.0	–
Synthetic staple fibre	3.6	4.2	4.2	8.3	11.3	14.9	17.7	23.3	26.3	31.0	36.0	39.1	51.5	76.8	81.4	111.0	123.0	134.0	145.0	–

Source: F.W. Rushing, 'Soviet Chemical Industry: a Modern Growth Sector' in *Soviet Economy in a New Perspective* (JEC, 1976). Updated for 1975, 1976, Narodnoye Khoziaistvo v 1976 godu; for 1977, Nar. Khoz. 1977.

Table 4

Expansion of capacity due to reconstruction or construction of new complexes and its share in total production

	1966-1970				1971-1975				1976		
Products and quantities	Total[1] added capacity	Average[1] annual increment	Total[2] produced 1966-70	Total added capacity as % of total produced	Total[1] added capacity	Average[1] annual increment	Total[2] produced 1971-75	Total added capacity as % of total produced	Total[1] added capacity	Total[2] produced 1976	Total added capacity as % of total produced
Mineral fertilisers (in terms of product), millions tons	33.2	6.6	220.7	15.0	38.0	7.6	370.4	10.3	7.3	92.2	7.9
Sulphuric acid, millions tons	4.2	0.8	52.9	7.9	8.6	1.7	76.6	11.2	2.6	20.0	13.0
Soda ash, thousands tons	1,221	244	16,575	7.3	1,038	208	21,197	4.9	246	4,842	5.1
Synthetic fibres and filament yarns, thousands tons	151.2	30.2	2730.0	5.5	349.4	69.9	4094.5	8.5	31.1	1,020	3.0
Synthetic resins and plastics, thousands tons	706	141	6501	10.9	981	196	11,561	8.5	308	3,058	10.1
Tyres, millions units	8.3	1.7	156.3	5.3	12.9	2.6	215.9	6.0	3.9	54.5	7.2

Sources: 1 *Vestnik Statistiki*, 1977, no.9, p.82.
2 Table 3.

'Chemicalisation' constitutes one of the most important pillars of the increasingly prominent strategy of intensive economic growth to which the USSR (and other socialist countries) has been turning since Stalin's death. It implies an expansion throughout the economy of the use of new materials, production of which is highly R and D intensive, but which when used in industry, agriculture and construction bring about vast savings attributable to the replacement of expensive traditional materials and to the labour savings connected with their use. Calculations worked out by the Soviet planners may give us some idea of the benefits that the use of synthetics in the economy brings. According to L. Kostandov, the All-Union Minister of the Chemical Industry, in 1980 the Soviet industry will utilise 7.8 tons of plastics for every RbL 1 mil. worth of industrial production as opposed to 5.5 tons in 1975. The share of chemical fibres in total output of textile raw materials will rise from 22 per cent in 1975 to 28 per cent in 1980. As a result of the increase in utilisation of plastics during this period the USSR will save 0.9 billion man-hours, RbL 2 bil. in prime costs, and RbL 1.1 bil. in capital investment. The growth of utilisation of chemical fibres in the same period will bring about the saving of RbL 1.3 bil. in miscellaneous expenditure and RbL 5.9 bil. in capital investment. At the same time the country will save 2.8 million tons of ferrous metals, 245 thousand tons of non-ferrous metals and some 5 million tons of timber. The increment in the supply of chemical fertilisers to agriculture to be registered during this period should result in further savings of about RbL 6.4 billion.[5]

These calculations may be interpreted in terms of an effort by the minister to illustrate the advantages of intensification of the Soviet economy which in this particular case is embodied in 'chemicalisation'. As will be shown in the subsequent discussion, a substantial part of the success of the strategy of implementing intensification via 'chemicalisation' is attributable to the policy of large-scale importation of Western chemical equipment to which the planners resorted (we have seen that the chemical industry belongs among those branches the technological development of which progressed least in the USSR, in part owing to the fact that it is relatively highly R and D intensive). If 'chemicalisation' may be said to be import-dependent then the same may be said about the intensification of the socialist economies in general. In chapter 1 we have suggested that after the termination of NEP the Soviet Union continued to reply on capitalist assistance in the sphere of science and technology. The increased emphasis on intensive methods of growth imply a reversion to the principles of NEP, this time on a higher level of Soviet economic development, and the role of Western capitalism in stimulating further progress of the socialist economies will grow in proportion to their dependence on the intensive factors. This will have far-reaching economic and political implications on both European and global levels. Above all, there now arises the question of whether this new opening of the Soviet system and its renewed policy of co-existence, compromise and co-operation with the West constitutes a permanent feature of the industrial strategy adopted in the USSR, or whether the USSR (and the other socialist countries) will subsequently gradually revert to the isolation characteristic of Stalin's years. All evidence would suggest that the latter course is highly unlikely because of the very heavy economic, and ultimately political, price the Soviet Union would have to pay should it wish to pursue it (see the conclusions in chapter 6).

'Chemicalisation' should also be seen as a part and parcel of the response on the part of the post-Stalin leadership to the technological gap that exists between East and West. The 'thaw' that followed in the years after the death of the dictator marked one of the most exciting periods in Soviet history. There was a renewed belief that communism, purified of the excesses of Stalinism, could after all go ahead to prove its superiority over Western capitalism, and on the industrial and technological front there were numerous signs that the USSR was entering a new glorious era which, as Khrushchev promised, would culminate in the institution of a full-blown communism and material abundance within a couple of decades. The technological reorientations which the post-Stalin leadership undertook were numerous: the change to oil and electricity traction on the railways, and from hydro- to thermal generation of electricity; the USSR began to modernise its food and textile industries and Soviet agriculture received a higher priority after the years of neglect by being re-equipped with modern engineering products and supplied with increasing quantities of chemical fertilisers (see chapter 4). The Soviet Union opened its 'Virgin Lands' and accomplished such technological feats as commissioning its first nuclear electricity generator and launching the first ever space satellite. 'Chemicalisation' itself had to wait until 1958 for its official blessing. This delay may in part be due to the well-entrenched traditional emphasis on coal as the favoured source of energy and organic chemistry feedstock,[6] but at the same time it must be pointed out that a major strategy of industrial change, augmented by massive imports of Western equipment, and the corresponding revolutionary transformation of the Soviet economy, could hardly be implemented while Stalin was still alive or even during the period immediately following his death. Five years elapsed between his death and the launching of 'chemicalisation', which testifies to the rigidity and motionlessness which permeated the whole system during the last years of his rule. Five years were needed by the Soviet Union before these new hopes and confidence could find their expression in its

Seven Year Plan of 1958.

B The Soviet inorganic chemical industry

Before looking at the two main pillars of 'chemicalisation' – petrochemicals and chemical fertilisers – it is necessary briefly to discuss inorganic chemical production in the USSR. This industry represents a more traditional and well-established branch of the chemical industry because the processes it uses as well as the products it turns out have been known for a relatively long time. Thus it can be concluded that today the focus of innovative activity is decisively outside the boundaries of inorganic chemistry. It is in organic chemistry that rapid technological advance has taken place in the past decades and that further progress in the future is expected to materialise. However, this does not mean that the inorganic chemical industry has lost its importance. The products falling within it constitute the foundation on which a major part of the modern chemical industry stands. The key products belonging to this branch are acids such as sulphuric, nitric or hydrochloric acid, ammonia and the group of alkali products – soda ash, caustic soda and chlorine. These products find an enormous variety of application in all branches of the chemical industry and in industry and services generally. For example, the acids are essential in the fertiliser industry or in the production of explosives; the alkali are necessary for the manufacture of rayon, pulp and paper, soap, glass, alumina and a number of other important products. They also find wide application in metallurgy, the food industry, oil refining, sanitation and waste neutralisation. Often the manufacture of well-known synthetic materials would be impossible without the availability of these inorganic substances (e.g. chlorine in the case of PVC).

As opposed to the other branches of chemical production the Soviet Union has always performed relatively well in terms of total output of most of these products. The USSR produced some 19 million tons of sulphuric acid in 1975 and the growth of output of this product between 1950 and 1970 was the highest of all industrial nations. It is the second largest producer of ammonia and the world's largest producer of soda ash. As regards other alkali, it can be noted that the USSR is the second largest producer of caustic soda and chlorine in Europe (next to the Federal Republic of Germany).

There were about 40 sulphuric acid plants before the revolution with a total output of about 121,000 tons per annum (tpa).[7] Half of these plants used the Russian Tenteleev contact process (which at the time was also used in the West). The remaining plants were based on the Grillo-Schroder or Manheim process. The Russians have successfully

Main Inorganic Chemical Plant Locations

scaled up this technology with the result that dependence on imported equipment in this particular product has been minimal. In recent years, however, the Soviet Union has begun to implement a major modernisation programme of the industry as part of the build-up of fertiliser capacity. The main feature of this programme is the increase in the size of plants and the considerable degree of reliance on relatively well-developed Polish sulphuric acid technology. In 1965 new plants had a capacity of 120,000 tpa whereas the plants commissioned in the mid-1970s reached the capacity of 450,000 tpa. The most recently ordered twelve Polish plants will have a capacity of up to 500,000 tpa. (By 1976 the USSR had ordered altogether 36 plants from Poland.)

There was hardly any nitrogen industry in pre-revolutionary Russia, which is surprising in view of its strategic importance (nitric acid and ammonia are used in the manufacture of explosives). But even after World War I there were only eight plants producing about 360 tpa of nitric acid. In the early 1920s Soviet officials surveyed the existing foreign technology for the manufacture of synthetic ammonia and nitric acid and decided to buy the Italian Casale process (ammonia) and the Du-Pont process (nitric acid). Subsequently the USSR also acquired from these companies processes for other related products. Western plants were built in the early 1930s as part of the programme of constructing chemical combines in several industrial centres (see below). The most dramatic build-up of this industry, however, has taken place since the launching of 'chemicalisation' in connection with the creation of the massive chemical fertiliser base. This aspect of 'chemicalisation' is discussed in the next chapter.

There were three soda ash plants in Russia before the revolution using the Solvay, Hemmingham and Le-Blanc processes. In 1930 the country produced about 55,000 tons of caustic soda. Expansion of the industry took place in the 1930s when the Soviet Union engaged a number of Western companies to build chemical combines and by implication the basis of the Soviet chemical industry. The most important of these combines are at the following locations: Berezniki-Solikamsk, Novomoskovsk (formerly Bobriki and Stalinogorsk), and Dzerzhinsk (the 'Kalinin' combinat, Gorkii oblast). The companies involved were Power Gas Ltd., Westvaco Chlorine Products, Inc., Nitrogen Engineering Corporation, Stockholms Superfosfat Fabriks Aktiebòlaget and others. They were responsible for supplying and erecting the equipment for the manufacture of a number of products including sulphuric and nitric acid, ammonia, chlorine, caustic soda, methanol and nitrogenous fertilisers. Co-operation with these companies was discontinued at the end of the 1930s after a series of disputes and breaches of contract on the part of the Soviet Union. More equipment, however, found its way into the USSR after World War II when as part of the reparation payments the Russians dismantled the German chemical complexes belonging to Deutsches Solvay Werke, Billiter, and Goldschmidt.

In the early part of this century the manufacture of caustic soda underwent an important technological change which consisted of replacing the old ammonia-soda process by electrolysis of salt brine solution (with chlorine as a by-product). The USSR acquired this progressive technology from the US company Westvaco Chlorine Products, Inc., and in the early 1960s the Russians purchased more of this equipment. Subsequently they successfully scaled-up and replicated this technology with the result that the Soviet-manufactured electrolytic cells are thought to be to a large extent replicas of the Hooker cell. Soviet engineers are at present working on a method incorporating the advantages of both of the two major types of electrolytic cells, the diaphragm and the mercury cell: the brine from the diaphragm cell is taken for further electrolysis in the mercury cell. In doing so, however, the Soviet Union merely follows the West since similar developments have taken place there.

Soda ash has traditionally been manufactured by the Solvay ammonia-soda process which was used in pre-revolutionary Russian factories. In recent years efforts have been made to replace it by a relatively new method based on the use of nepheline as raw material. Nepheline is mined on the Kola peninsula together with apatite and is used for the production of alumina. The technique yields cement and soda ash as a by-product, and given the high demand for cement and alumina the method has proved to be relatively cheap.[8] In 1975 about 16 per cent of soda ash produced in the USSR was based on this method.

Although recent development of the chlor-alkali industry is largely due to indigenous efforts, consisting of scaling-up and piecemeal improvements, sporadic purchases of foreign technology did take place. The scale of these purchases, however, does not in any way approach the massive proportions registered in the petrochemical industry. It may, therefore, be concluded that in this technologically well established branch of chemical production the Soviet Union is performing relatively well and has achieved virtual self-sufficiency. Table 5 illustrates this point by documenting the relatively insignificant recent purchases of alkali technology.

Table 5
Recent Soviet purchases of chlor-alkali equipment

Company	Location	Product	Capacity	Year of completion
Krebs et Cie	Novomoskovsk	Caustic soda chlorine	200,000 tpa	1962
Krebs-de-Nora/ Technimont	Sterlitamak	Chlorine and caustic soda	140,000 tpa	1977
Krebs-de-Nora/ Technimont	Sterlitamak	Salt crystallisation plant	270,000 tpa	1977
Krebs-de-Nora	Komsomolsk	Chlorine	6,270 tpa	–

Note: Chlorine equipment is also imported as part of the purchases of petrochemical technology in connection with the construction of PVC plants, see, for example, Zima in Siberia (see Appendix B).

Source: Based on the reports by *European Chemical News* and *Chemical Age*.

C Soviet petrochemicals and downstream products

The creation of the industrial base capable of turning out enormous quantities of petrochemicals and their end products constitutes the hallmark of Soviet 'chemicalisation'. The products belonging to this group are very numerous but they can be divided into four groups:

a) heavy organic chemicals obtained by catalytic cracking of hydrocarbons. Most important of these are the aliphatic substances which include: ethylene, propylene, acetylene and butadiene. Another major group are the aromatic fractions such as benzene, xylenes, phenol, naphthalene and pthalic anhydride. These basic materials form feedstock for the production of downstream products the main groups of which are:

b) plastics and resins, which include the polymers: polyethylene, polypropylene, polystyrene and polyvinylchloride (PVC); and polycondensates: synthetic resins such as phenol-formaldehyde and urea resins.

c) chemical fibres; the types produced in the Soviet Union are: polyamides (*Kapron*), polyester (*Lavsan*), polyacrylonitrile (*Nitron*), and apart from these the USSR also produces some types of specialised fibres such as polypropylene fibres, florofibres, PVC fibres and polyvinyl alcohol fibres.

d) synthetic rubber; the main types produced in the USSR are the following: stereoregular, styrene-butadiene, butadiene-metylstyrene, polybutadiene, nitrile and chloroprene. The more special-purpose types include neoprene, silicone, polysulphide and others.

Not enough is known about the output structure of the Soviet heavy organic chemicals[9] (group (a)) and consequently we shall begin by looking at the production patterns of the two main groups – plastics and resins and chemical fibres – and supplement this analysis by a brief outline of the Soviet synthetic rubber production.

In his study of the technological level of the Soviet chemical industry Amann finds that the production pattern of plastics, as well as of chemical fibres, testifies to the existence of a substantial lag behind major industrialised nations because the Soviet Union's share of technologically inferior materials stubbornly continues to be relatively very high. The USSR is the world's second largest producer of the more traditional phenolformaldehyde resins while it lags well behind the leading Western nations in the production of technologically more progressive materials such as polystyrene, PVC, polyethylene and polypropylene. Table 6 presents the output data of various types of plastics and resins. In the West the output of the latter group (polymerised plastics) tends to be about half, or less, of that of the former group (polycondensation plastics) but in the Soviet Union this proportion is approximately reversed.[10] The Soviet planners are aware that the technologically most advanced products bring about the greatest economic returns and efforts are made to correct this situation. According to Kostandov the target of the current plan is to increase the share in the total output of the technologically most progressive polymers from 35.2 per cent in 1975 to 50 per cent in 1980.[11] As we shall see, Western technology plays a key role in this qualitative (as well as quantitative) improvement of the Soviet chemical output because the equipment for producing polymers has prominently figured on the Soviet 'shopping list'.

In the case of fibres the story is much the same. Amann shows that the Soviet Union lags considerably behind the West both in terms of *per capita* level of output and in terms of the relative proportions of the technologically more sophisticated fibres. In the West a decisive switch occurred away from the man-made fibres based on cellulose in favour of synthetic fibres. If in the West by the early 1970s the synthetic fibres accounted for well over half of the total output, in the Soviet Union the share of the synthetic fibres amounted to less than one third of the total. Table 7 shows that this disproportion is gradually being

Table 6
Production of plastics and synthetic resins (000 tons)

	1970	1975	1975 %
Total	1,673	2,800*	100
Polymerised	585.55	985.60	35.2
of which			
polyethylene	267.68	420.00	15.0
polypropylene	10.03	11.20	0.4
polystyrene and copolymers	83.65	142.80	5.1
PVC and copolymers	160.60	330.40	11.8
Polycondensates	846.53	1,318.80	47.1
of which			
phenolformaldehyde resins	142.20	204.40	7.3
urea resins	347.98	613.20	21.9
Fibre forming	240.91	495.60	17.7

*The USSR intends to double the production of these materials by reaching the target of 5.7 million tons in 1980.

Table 7
Production of fibres in the USSR (000 tons)

		of which							
			of which				of which		
Year	Total Man-made	Cellulosic	Rayon filament (excl. rayon filament for cord)	Rayon filament for cord	Rayon staple fibre	Synthetic	Filament (excl. filament for cord and technical purposes)	Filament for cord and technical purposes	Staple fibre
1970	623	456	93	129	234	167	39.5	76	51.4
1971	676	473	98	132	243	203	42	90	71
1972	746	507	105	140	262	239	48	109.3	81.4
1973	830	543	110	149	284	287	52	124	111
1974	887	569	111	156	302	318	54	141	123
1975	955	590.2	n.a.	n.a.	n.a.	364.8	n.a.	n.a.	n.a.

Source: *Chemical Industry of East Europe 1975-80* (IPC Business Press)

Table 8
Production of man made fibres (000 tons)

	1970	1975	1980 Plan
Total	623.0	955.0	1450.0–1500.0
Artificial	456.7	590.2	700.0
of which			
viscose	400.6	526.2	–
acetate	31.8	46.8	–
Synthetics	166.3	364.8	745.0
of which			
polyamide (capron)	133.3	233.0	430*
polyester (lavsan)	23.7	75.4	210*
polyacrylonitrile (nitron)	8.1	53.5	100*
others	1.2	2.9	5*

*Estimated

Source: L.A. Kostandov, Minister of Chemical Industry. Quoted by *Chemical Industry of East Europe 1975-80* (IPC Business Press)

corrected, although Amann argues that the continued expansion of output of the cellulosic fibres at a time when the synthetic sector has been undergoing rapid development is a manifestation of persistent conservatism. However, as in the case of plastics and resins, the planners seem to be well aware of the advantages of the technologically most progressive synthetic fibres. The total target for 1980 is 1,460 thousand tons which implies a 1.5 times increase over the 1975 level. At the same time the output of synthetic fibres is scheduled to rise by the factor of 2 while cellulose-based materials will rise only 1.5 times. This means that the synthetics' share will increase from 38 per cent in 1975 to 52 per cent by the end of the decade.[12] These intentions are also illustrated in table 8. Again, any such technological updating of the production structure will be accomplished through purchases of Western technology.

The Soviet Union is secretive about its synthetic rubber industry but it happens to be a branch in which it has traditionally performed relatively well. According to Sutton the USSR may rightly be regarded as a pioneer in the development of this strategically important material. First discoveries were made in pre-revolutionary Russia and further progress was made in the 1920s, and although the first commercial production started in the United States the USSR did in the mid-1930s succeed in independently developing its own process yielding the sodium-butadiene (SK-B) type of rubber. Since Soviet statistics present the data relating to the output of tyres, one way of estimating the production of synthetic rubber is to assume that about 70 per cent of Soviet rubber is used in the manufacture of tyres. On the basis of this method Amann estimates[13] that in 1972 the USSR produced some 982.8 thousand tons and that it has been one of the world's leading producers, surpassed only by the United States and strongly challenged by Japan. The International Rubber Study Group estimated that in 1975 the Soviet rubber output reached 1,325 thousand tons. During the current plan period production is to increase by 40-60 per cent. The Soviet data on the structure of output are reproduced in table 9.

Table 9

Production structure, %	1970	1975
Total	100.0	100.0
Stereoregular	29.5	42.8
Styrene-butadiene and butadiene-methylstyrene	47.6	36.9
Polybutadiene rubber	9.2	5.2
Nitrile (acrylonitrile-butadiene)	2.5	2.9
Chloroprene	7.7	4.4
Others	3.5	7.8

Source: *Chemical Industry of East Europe 1975-80* (IPC Business Press)

D Soviet petrochemicals and Western technology

Dependence on co-operation with Western companies in the oil refining sector dates back into the 1920s and 1930s when the Russians engaged various oil companies to construct new refineries, or expand the existing ones at Baku, Tuapse, Grozny, Emba, as well as those situated along the Volga (the so called 'Second Baku'), e.g. at Ufa. Further Western technological assistance enabled the Soviet Union to build its refineries at Guriev, Omsk, Kuibyshev, Krasnovodsk, Sterlitamak and Moscow (see the map 'Refineries and petrochemical derivatives plant locations' in Appendix B). Sutton has shown that all subsequent refineries built by the USSR are based on the original Western designs.[14]

At the start of 'chemicalisation' these traditional oil refining regions accounted for about three-quarters of the total output. Since then a process of decentralisation has taken place with new refineries being constructed in Eastern and Western Siberia and in the European part of the Soviet Union especially in the Ukraine and Byelorussia. Thus new refineries were built at the following locations: Atchinsk, Chardzhou, Lisichansk, Mozyr and others. Further expansion is presently taking place in many of these refining centres. Whereas in the early period of

Soviet industrialisation the refineries were geared mainly to the production of fuels, today, under the imperatives of 'chemicalisation', the Western companies are engaged in the programme of constructing a petrochemical base. The degree of Soviet dependence on Western technological assistance throughout the period of 'chemicalisation' has been virtually total. As a general rule it can be said that the greater the technological sophistication of a given product in terms of complexity of the process on which its output is based, the greater the degree of dependence on Western equipment. Thus virtually all major types of plastics and fibres in the USSR are produced by imported plants. Although Amann's study of the Soviet chemical industry is only indirectly concerned with the issue of Western technological assistance, it shows, nevertheless, that even when the Soviet Union succeeded in developing its own process for the production of some type of plastics or fibres these initial advances were followed by imports of Western plants, which suggests that the processes developed in the West were ultimately judged by the Russians to be technologically superior and more economical.[15]

Even as regards synthetic rubber the further development of this group of products in the USSR proved to be disappointing. The Soviet rubber industry has failed to develop a greater variety in the types of rubber produced with the consequence that the range of properties of rubber produced in the USSR was soon found insufficient in regard to specific industrial needs. Thus after a relatively early start the Soviet Union began to lag behind the other industrial powers in synthetic rubber technology. This lag was partially closed by World War II Lend-Lease arrangements according to which complete plants were acquired by the USSR in order to update the level of its industry and to match the needs of its war effort. After the war the Soviet Union dismantled the IG Farben complexes for the production of the German Buna-S type of synthetic rubber.[16] Since the war the Soviet synthetic rubber R and D has benefited enormously from the continued co-operation with the GDR. However, Western companies continue to be involved (for example, in Yerevan and Tobolsk: see Appendix B). Apart from this the Soviet Union has purchased a number of tyre factories as table 10 shows. The more recent acquisitions of tyre plants are largely a part of the programme of building a more up-to-date automotive industry with the help of Fiat, Renault and other Western companies. Apart from tyre plants the Soviet Union also purchased equipment for the manufacture of other rubber components related to this programme and, of course, the production of Soviet tyres would be impossible without Western assistance in producing the special viscose rayon or nylon tyre cords. These plants have been supplied by Chatillon (1963, 1966) and Courtaulds (1964) and erected at Balakovo.

Table 10
Soviet acquisitions of tyre plants

Year	Supplier of plants and equipment	Location
1937[1]	Seiberling Rubber Co. Inc./Francis Shaw & Co. Ltd.	–
1945[1]	Ford Motor Company	–
1946[1]	Deka-Werke (German reparations)	–
1946[1]	Manchu Rubber Co. (Manchurian reparations)	–
1957[1]	'Rustyfa' consortium	Dnepropetrovsk
1957[1]	Chatillon	–
1960[1]	Simon Handling Engineers Ltd.	Krasnoyarsk
1960[2]	Pirelli	Yaroslavl
1970[2]	Pirelli	Balakovo
1977[2]	Pirelli	Kama River
1977[2]	Pirelli	Moscow
1977[2]	Krupp	Minsk

Sources: 1 Reprinted from *Western Technology and Soviet Economic Development* (vol.3, p,160) by Antony Sutton, with the permission of the publishers, Hoover Institution Press. Copyright 1968 by the Board of Trustees of the Leland Stanford Junior University.
2 Based on the reports by *European Chemical News* and *Chemical Age*.

On the basis of a survey of the press we have assembled all publicly available information about the contracts between the Soviet Union and its Western partners participating in the construction of the Soviet petrochemical industry. Since the table containing this information is too large to be included in this chapter, it is reproduced together with the accompanying map 'Refineries and petrochemical derivatives plant locations' in the Appendix B to which the reader is referred. Tables 11 and 12 supplement the information given in the Appendix. The tables show that approximately 50 per cent of all contracts signed since the inception of 'chemicalisation' belong to the category of petrochemicals and downstream products. The next largest group is that of chemical fertilisers. At the beginning of the period most deliveries came from the United Kingdom followed in order of magnitude by France, Italy, the Federal Republic of Germany, Japan, the United States, and other West and North European countries. Today the largest contractor is the FRG which since the launching of its *Ost-Politik* in the late 1960s has become the leading trader with Comecon countries.

Table 11
Import of Western chemical plant and equipment 1960-75 by type and number of contracts

	Number of contracts				
	1960-65	1966-70	1971-75	Total	Per cent of total
Fertilisers	27	12	11.8	50.8	16.0
Plastics	20	20	22	62	19.5
Fibres	16	8	13.2	37.2	11.7
Rubber	10	3	2	15	4.7
Rubber goods	9	5	8	22	6.9
Petrochemicals	8	5	5.6	18.6	5.8
Other goods	13	8	10.2	31.2	9.8
Inorganic	5	2	2.2	9.2	2.9
Household	2	1	1	4	1.3
Paper	3	13	8	24	7.5
Pharmaceuticals	0	6	5	11	3.5
Pesticides	3	4	0	7	2.2
Automation	2	0	6	8	2.5
Other	3	6	9	18	5.7
Total	121	93	104	318	100.0

Source: *Chemical Age* Survey: 9 June 1978 (Morgan Grampian Ltd.).

Table 12
Chemical equipment imports from the West by country and number of contracts 1960-1975

	Number of contracts				
	1960-65	1966-70	1971-75	Total	Per cent of total
Austria	3	0	1	4	1.3
Belgium	4	1	0	5	1.6
Denmark	1	0	0	1	0.3
Finland	1	7	1	9	2.8
France	10	18	16	44	13.8
Italy	11	10	17	38	11.9
Japan	7	22	11	40	12.6
Netherlands	6	4	0	10	3.1
Sweden	2	2	3	7	2.2
Switzerland	3	2	0	5	1.6
United Kingdom	49	18	25	92	28.9
United States	3	2	7	12	3.8
West Germany	21	7	23	51	16.0
Total	121	93	104	310	100.0

Source: *Chemical Age* Survey: 9 June 1978 (Morgan Grampian Ltd.).

The share of chemical equipment shipments by value provides a slightly different order. Between 1971 and 1975 the USSR imported RbL 384 million from Western Germany. This represents 31 per cent of total imports from the West. Japan followed with RbL 256 million, France with RbL 193 million, Italy with RbL 143 million and the UK with RbL 107 million.[17]

We have seen that the main reason why the USSR has found it necessary to resort to this large-scale programme of chemical equipment acquisitions is that it is unable to generate its own technology to satisfy its requirements. In many instances the Soviet Union would simply not be able to produce some products at all were it not able to purchase the relevant plant abroad. But another factor has to be taken into consideration, namely the fact that the quality of Western based products is often superior to that produced on the basis of comparable Soviet or Comecon technology and that the Western plants as a rule have a much bigger capacity than the Soviet-designed plants. This enables the USSR to benefit from economies of scale. Apart from complete chemical plants the Soviet Union has recently shown increasing interest in highly advanced technology such as computer control equipment and whole ethylene pipeline transport systems. Soviet dependence on imports of Western technology is likely to continue in the foreseeable future because the nature of this equipment is such that it cannot be replicated and/or scaled up in the traditional Soviet manner. The processes and equipment involved are too complex to be mastered through 'reverse engineering', and often the modern plants utilise metal piping, valves, containers, etc., which the Soviet steel industry, despite its relatively high technological level, cannot produce.

An eloquent example of this weakness is the degree of Soviet dependence on imports of the large 1.200 mm and 1.422 mm diameter steel pipes which the Soviet industry apparently cannot produce and the lack of which would considerably slow down the progress of 'chemicalisation'. In the current Five Year Plan the USSR plans to build about 9,300 miles of crude oil pipelines and 22,500 miles of gas pipelines. The corresponding figures for the previous plan period were 12,100 and 20,100 respectively. Between 1970 and 1977 the Soviet Union bought about 9 million tons of steel pipes mostly from West Germany, Italy, Japan, France and from the Comecon countries.[18]

It should also be pointed out that when engaging a Western firm to erect a particular piece of chemical equipment the planners may be assured that the job will be done within a relatively short space of time and that any unscheduled delays will be minimal (and probably caused by the Russians themselves). Competition in the West for East European contracts is very high and the companies make sure that they do their best to satisfy their socialist customers. This means that the Soviet

Union not only buys superior technology but also superior services and management, both of which it cannot produce domestically. The construction of Western plants takes about two to four years and may thus conveniently be fitted into a Five Year Plan's time span. Russian plants may take about twice as long to construct.

E Specific characteristics of East-West technology transfer

Finally, a few words are necessary about the mechanics of technology transfer itself. There are basically two forms in which this process takes place anywhere in the world. An established and technically competent firm may buy the right to use a particular patented process to improve its efficiency or extend its range of output. With the rights to the process the recipient probably buys the 'know-how' in the form of drawings, layouts, machinery lists, process details, control information, etc. Should the patent have run out, the firm may still buy 'know-how' in the same way. A firm with less knowledge of the technology involved may need to buy the 'do-how' as well as the 'know-how'. In this case it is more likely to buy the technology embodied in a complete installation – a 'package' or 'turn-key plant'. Such a firm then buys the authority to use the process with all the technical detail as above. Additionally, however, it may get its staff trained in the art of the process by the licenser and have the plant started up under supervision to ensure that it meets the specifications in the contract (this is known as 'guarantee trials'). The buyer may also include clauses covering the updating of the process with newer technology as it becomes available over a period of time.

The USSR and the other Comecon countries as a rule opt for the latter, more comprehensive form of technology transfer. But there are some additional specific features which lend East-West technology transfer its unique character. First, the Soviet Union is represented in such negotiations by a monopolistic state foreign trade organisation (Tekhmashimport in the case of the Soviet purchases of chemical equipment) which rather than pay licence fees prefers to buy the patent rights for its whole territory outright as a part of the overall cost of the whole deal. Since in the West patents are in most cases the property of individual companies rather than governments the bargaining of a company and of a monopolistic socialist buyer is often unequal. This brings us to the second important feature. A contracting firm, in order to win an order, must not only design the plant, procure the equipment, install the machinery[19] and negotiate the contract through the bureaucratic pitfalls of the Soviet system, it must in addition negotiate the necessary finance (obtain credit, etc.) and, as is increasingly the case with Comecon in recent years, take a specified quantity of the resulting product back in payment for thc installed plant (see chapter 5). The contractor then is required to market the product in the West either with the help of a partner or by its own marketing organisation. Western assistance, therefore, is frequently not confined to the mere transfer of technology in the conventional meaning of the concept. It also involves, as noted above, a transfer of financial resources and managerial skills.

A major difficulty faced by Western firms when dealing with Soviet organisations is that the former cannot be quite certain whether or not they are putting a fair price on their technology. Technology is one of the main assets of a company; it may have spent considerable financial and human resources on its development and naturally it would like to see it recouped. The transfer of technology anywhere in the world is the result of an investment decision on the part of the recipient, in our case the USSR, but it is very difficult to set a fair price for which a Western company should ask. A lump sum is usually determined in the course of negotiations, but this can hardly be a satisfactory method in view of the fact that the transferred technology may be exploited in replicated plants throughout the vast territory of the Soviet Union. Moreover, technology transfer business is highly competitive and consequently the price paid is normally as low as anywhere in the world.

There is also the question of security of patents in the CMEA. The USSR has it own system of rewarding inventors but, nevertheless, in 1965 it accepted the validity of the international patent system by acceding to the Paris Convention of 1884 for the Protection of Industrial Property. The security the system provides gives the inventor an incentive not to keep his technology secret, secrecy being the only alternative means of protecting it. The Comecon countries claim to be able to ensure the necessary standards of security for patents they wish to purchase from the West but important cases have occurred where such information supplied to one country has, after a period of time, beeen transferred to another. Although such practice has become extremely rare in recent years as the Comecon countries have a vital interest in promoting their good reputation, some causes of concern have not completely disappeared. In the early years of Comecon's existence the member countries supplied each other with available technical documentation free of charge, which substantially contributed to the boosting of production throughout the region. This practical expression of the injunctions of communism[20] is no longer practised as R and D became more and more resource-demanding. Nevertheless, the possibility of a continued existence of some measure of mutual solidarity among the socialist countries in respect to acquisitions of technical knowledge from the capitalist West cannot altogether be excluded.

Lastly, the subject of East-West transfer of technology raises the question of whether it should be pursued at all bearing in mind the indisputable fact that the West has been supplying high level technology to countries whose social, political and economic system is radically different from the political and economic practice of the West and who, moreover, claim that East-West *détente* does not imply a renunciation of the international class struggle and indeed whose declared objective is to destroy Western capitalism and pluralistic political systems. East-West technology transfer, therefore, has far-reaching security implications, and the tenacity with which it has been practised ever since the origins of the USSR, in spite of the difficulties and obvious security risks, is perhaps one of its most bizzare characteristics. (These themes are discussed in the conclusions, chapter 6.)

Notes

1 R. Hutchings, *Soviet Science, Technology, Design*, pp 186-7.
2 R. Amann, 'The (Soviet) Chemical Industry: its Level of Modernity and Technological Sophistication' in Amann, Cooper and Davies, op.cit. (chapter 2, note 12), p.250.
3 F.W. Rushing, 'Soviet Chemical Industry: a Modern Growth Sector' in *Soviet Economy in a New Perspective*, JEC, 1976, p.541.
4 Ibid.
5 L.A. Kostandov, 'Khimicheskaya promyshlennost' v desyatoy piatiletke', *Zhurnal vsesoyuznovo khimicheskovo obshchestva imeni Mendeleyeva*, no.2, 1977, p.129.
6 R. Hutchings, *Soviet Science, Technology, Design*, p.184.
7 The historical aspect of the following outline relies on the information provided by A.C. Sutton, op.cit. (chapter 2, note 5), vol.2, chapter 7.
8 T. Shabad, 'Raw Material Problems of the Soviet Aluminium Industry' in *Soviet Economy in a New Perspective*, JEC, 1976, p.666.
9 Data on the Soviet and CMEA production of these chemicals appeared in C. Rajana, *The Chemical and Petrochemical Industries of Russia and Eastern Europe 1960-1980*, Sussex University Press, 1975, Appendix. Their origin, however, is not indicated.
10 R. Amann, 'The Chemical Industry: Technological Sophistication' in Amann, Cooper and Davies, op.cit. (chapter 2, note 12), p.253.
11 L.A. Kostandov, loc.cit. (above, note 5), p.126.
12 Ibid.
13 R. Amann, 'The Chemical Industry: Technological Sophistication'

in Amann, Cooper and Davies, op.cit. (chapter 2, note 12), p.256.
14 Sutton, op.cit. (chapter 2, note 5), vol.3, p.136.
15 Amann, op.cit. (chapter 2, note 12), p.284.
16 Sutton, op.cit. (chapter 2, note 5), vol.3, p.154.
17 O.K. Denisov, 'Vneshneekonomicheskie sviazy sssr s kapitalisticheskimi stranami v oblasti khimicheskoy promyshlennosti', *Khimicheskaya promyshlennost* no.12, 1977, p.38.
18 *The Oil and Gas Journal*, October 1977, p.110.
19 The transfer of chemical technology faces unaccustomed difficulties. The Russians normally insist on themselves erecting the buildings, which are rarely available on time. These are built to the '6 metre norm'. This is often badly suited to the installed process requirements. It is furthermore often unnecessarily costly in terms of wasted space in gangways and around machines. Because the availability of labour is not directly under the control of the engineer responsible for the plant, process working is often erratic. Services are likewise unpredictable. In short, a vastly greater effort has to be made in USSR and many East European countries than elsewhere.
20 M. Kaser, *Comecon*, OUP, 1967, p.153.

4 Soviet chemical fertilisers and East-West transfer of technology

The creation of a modern chemical fertilisers industry constitutes the second most important aspect of 'chemicalisation'. As in petrochemicals the implementation of this programme has depended on imports of Western plants and equipment. This chapter follows both thematically and structurally chapter 3: first we shall discuss the general attitudes of Soviet leaders to the questions of developmental strategies as well as the main changes the chemical fertilisers industry has undergone in recent years, together with an outline of the paths the planners intend to pursue in the future. Our next step will consist of illustrating the extent of imports of Western plants against this background in order to gain a picture of the role that foreign technology has been assigned in the overall development of the industry.

A Chemical fertilisers and Soviet agricultural policy

It is to some extent a paradox that Khrushchev, who was far sighted enough to recognise the importance of a modern chemical industry and who, moreover, was deeply involved in the problems of agriculture, was relatively slow in fully appreciating the economic potential of chemical fertilisers. Although the foundation of the industry (in its modern sense) is firmly associated with his name, his record in agricultural policy is marked above all by his commitment to the extensive strategy of increasing the agricultural output by expanding the size of the sown area by cultivation of the so called Virgin Lands areas in the Far East, the Urals and the Volga regions, Kazakhstan and Siberia. It would probably be incorrect to suggest that the construction of the chemical fertilisers industry was as a result neglected, because impressive strides during the years of Khrushchev's leadership were undertaken, but his political commitment to campaigns such as the Virgin Lands cultivation certainly meant that the industry did not receive the degree of priority it might otherwise have got.[1] The natural fertility of the virgin and long unused soil was supposed to postpone the need for artificial inputs, and scientific advice by experts was ignored. It was not until 1963-64 that Khrushchev admitted that he had got his priorities wrong, but by that time he had been removed from his post of General Secretary of the Communist Party. Nonetheless, his Seven Year Plan foresaw a rapid growth of fertilisers output from 12.4 million tons in 1958 to 35 million tons in 1965, a growth rate of 16 per cent per annum during the period. As table 3 shows, the actual performance failed to match these expectations; only 31.3 million tons were produced in 1965.

With Khrushchev's departure from the scene, chemical fertilisers began to play a pivotal role in the Soviet leadership's agricultural development strategy. In March 1965 various resolutions were passed by the Central Committee which put an official stamp on the change of emphasis in favour of a new agricultural policy which broadly speaking rests on implementation of the following measures:

a) emphasis on 'intensive' methods of stimulating output increase by application of fertilisers,

b) a land improvement programme of draining humid areas and irrigating dry areas,

c) increased investment in agriculture, agriculture supporting industries and rural infrastructure.

Since 1965 investment in Soviet agriculture has grown about two-thirds faster than in the rest of the economy with the result that agriculture today takes about 27 per cent of all state capital investment. A broader concept of agricultural investment, which includes a variety of agricultural supporting industries, causes this share to reach up to about 34 per cent of the total.[2]

The overall objective of the post-Khrushchev agricultural policy, initiated in 1965-66 but being gradually more strongly pronounced in the 1970s, has consisted of the creation of massive agro-industrial complexes which are thought to be capable of facilitating a significant improvement in agricultural performance without having to rely on various *ad hoc* reorganisations and spectacular campaigns characteristic of the Khrushchev era and without being forced even to contemplate fundamental and ideologically unacceptable reforms, such as at least a limited abolition of collectivisation. Tables 13A and B illustrate the extent of these efforts both in land improvement and in provision of the necessary agricultural industrial infrastructure.

The essence of this policy is exemplified in the concentration of these resources on the so-called Non-Black-Earth-Zone of the Russian Soviet Federative Socialist Republic. This area, which extends from the Baltic to the Urals, contains over 50,000,000 Ha, some 32,000,000 Ha of which is arable land. It amounts to more than one-fifth of the total agricultural land in the Soviet Union and nearly a quarter of the popu-

Table 13A
Efforts to improve cropland, planned and actual[1]

	Annual average							Annual average		
	1966-70 plan	1966-70 actual	1971	1972	1973	1974	1975	1971-75 plan	1971-75 actual	1976-80 plan
Mineral fertiliser, deliveries to agriculture:[2]										
Million tons, standard units	41.4	37.0	50.5	54.8	60.0	65.9	75.4	60.6	61.3	93.4
Per cent increase	15.2	11.0	10.7	8.4	9.5	9.8	14.5	10.4	10.6	9.7
Area limed:										
Million hectares	6.0	4.5	5.2	5.5	5.9[4]	6.4[4]	7.0[4]	6.4	6.0	8.0-10.0
Per cent increase	25.5[1]	11.7	3.3	6.4	8.3	8.3	8.3	8.4[1]	7.0	4.5-12.1[3]
Gross addition of irrigated land:										
Thousand hectares	550.0	360.0	515.0	784.0	960.0	1,090.0	1,250.0	800.0	920.0	980.0
Per cent increase	11.3[1]	– .4	33.4	52.2	22.4	13.5	14.7	25.5[3]	26.0	– 8.0[3]
Gross addition of drained land:										
Thousand hectares	1,250.0	782.0	834.0	837.0	905.0	815.0	1,017.0	1,000.0	882.0	940.0
Per cent increase	19.6[1]	2.9	2.3	.4	8.1	–9.9	24.8	6.9[3]	4.5	– 2.6[3]

1 Rates of growth calculated from unrounded data.
2 Includes feed additives.
3 Constant rates of growth derived from actual performance in the base year and planned total performance for the succeeding five-year period.
4 Estimated.

Source: D.W. Carey, 'Soviet Agriculture: Recent and Future Plans' in *Soviet Economy in a New Perspective* (JEC, 1976), pp 586-7.

Table 13B

Deliveries of machinery and equipment to agriculture, planned and actual

	Annual average							Annual average		
	1966-70 plan	1966-70 actual	1971	1972	1973	1974	1975	1971-75 plan	1971-75 actual	1976-80 plan
Deliveries to agriculture of:										
Tractors:										
Thousand units	358	293	313	313	323	348	370	340	333	380
Rate of growth[1]	13.7[2]	5.2	1.3	−0.1	3.3	7.7	6.4	3.2[2]	3.7	0.9[2]
Trucks:										
Thousand units	220	143	169	188	225	251	269	220	220	270
Rate of growth[1]	29.7[2]	10.7	8.2	10.7	19.9	11.6	7.4	11.6[2]	11.5	0.1[2]
Agricultural machinery:										
Billion rubles	2.2	1.8	2.5	2.7	3.1	3.5	3.8	3.1	3.1	4.6
Rate of growth[1]	11.3[2]	7.2	16.0	11.4	13.1	11.9	8.9	12.9[2]	12.2	6.6[2]
of which, combines:										
Thousand units	110	94	99	93	82	83	92	109	90	108

1

2

Source: D.W. Carey, 'Soviet Agriculture: Recent and Future Plans' in *Soviet Economy in a New Perspective* (JEC, 1976), pp 586-7.

lation resides there. (In fact, the NBEZ is the most densely populated area in the USSR.) As its name suggests, the zone does not have the best quality soil: it has large tracts of uneven and stony land with low natural fertility, high acidity and waterlogging.

The development of this land, however, has several advantages. It is closer to the centres of consumption of agricultural output and by the same token to the centres of agriculture-supporting industries including the fertilisers industry. The NBEZ has relatively frequent rainfall which makes fertilisers fully effective, and although drought does occur there it is relatively less susceptible to it than the other regions of the Soviet Union. All this has the combined effect that the relatively high investment required by the NBEZ brings adequate returns and is substantially more secure than elsewhere. The preferential treatment of the NBEZ was initiated in the mid-1960s and, as table 14 shows, its share of the total agricultural investment has been increasing over the current decade.

Table 14
USSR: resources invested in the
Non-Black-Earth-Zone of the RSFSR

	1971-75		1976-80 plan	
	USSR	NBEZ	USSR	NBEZ
Total gross fixed investment in agriculture:				
Billion rubles	131.5	19.5	171.7	35.0
Per cent share	100.0	14.8	100.0	20.4
Deliveries of:				
Tractors:				
Thousand units	1,687	287	1,900	380
Per cent share	100.0	17.2	100.0	20.0
Trucks:				
Thousand units	1,102	190	1,350	230
Per cent share	100.0	17.2	100.0	17.0
Grain combines:				
Thousand units	449	73	538	94
Per cent share	100.0	16.2	100.0	17.5
Fertiliser:				
Million tons	307	63	467	120
Per cent share	100.0	20.5	100.0	25.7

Source: D.W. Carey, 'Soviet Agriculture: Recent and Future Plans' in *Soviet Economy in a New Perspective* (JEC, 1976), pp 586-7.

The improvement of soil of a lesser quality rather than the spectacular Virgin Lands campaigns engineered by Khrushchev has been the main characteristic of agricultural policy in the Soviet Union under

Table 15

A. Deliveries of fertilisers to industrial crops and cereals (nutrient kg/ha)

Cotton			Sugar beet			Cereals (excluding maize)		
1974	1975	1976	1974	1975	1976	1974	1975	1976
367	391	393	299	399	459	40	42	47

B. Per cent of the total sown area to which fertilisers were applied

Cotton			Sugar beet			Cereals (excluding maize)		
1974	1975	1976	1974	1975	1976	1974	1975	1976
98	99.5	99.5	98	99.4	99.5	48	48	50

C. Share of Soviet crop land taken by cereals, fodder and industrial crops (millions of hectares)

Type of crop	1960	1973	1976
Cereals	122	128	128
Industrial	13	15	15
Fodder	58	63	66
Total	204	215	218

Sources: *Vestnik Statistiki*: 1975, no.5, p.92; 1976, no.5, p.95; 1977, no.4, p.94.

Brezhnev's leadership. The caution exhibited by this leadership with respect to agriculture goes hand in hand with the adoption of a similar attitude in the other areas of policy-making. In any case, there is today little prospect of significant gains in expanding crop land in the USSR. Between 1970 and 1975 the crop land increased by 5.3 per cent, but it must be noted that a large proportion of this increase merely compensates for the excessive land erosion and claims of the construction industry.

Another important feature of Brezhnev's agricultural policy has been an increasing emphasis on stimulating the volume of production of meat and dairy products in order to improve the quality of the Soviet population's diet. Chemical fertilisers came to play a major role in pursuing these aims. This represents a profound change of attitudes because even under Khrushchev the idea of applying fertilisers to grain cropland, but especially to fodder crops, was extremely slow in becoming acceptable. Calculations show that the value of output per unit of land under cereals is roughly equal to that of land under fodder. But industrial crops give about six times, and vegetables (including potatoes) about thirteen times, the value of cereals.[3] For this reason these crops traditionally received a higher priority in application of fertilisers than grain. The present leadership pays greater attention to the consumerist expectations of the population and chemical fertilisers are perceived as an integral part of the strategy to overcome the chronic weaknesses of Soviet agriculture and to provide sufficient quantity and quality of foodstuffs. If these efforts prove to be successful, the decision to construct a large chemical fertiliser industry could bring about important political dividends as well as hard currency savings because the USSR would not need to purchase grain abroad. However, as table 15(A) shows, the current situation still leaves much to be desired. Between 1974 and 1976 the grain cropland received about 8-10 times less chemical fertiliser per Ha than cotton and sugar beet. Table 15(B) indicates that in 1976 only 50 per cent of the total grain cropland in the USSR received any fertiliser at all, while table 15(C) shows that cereals are sown on about 55 per cent of the total cultivated land in the Soviet Union. Clearly the demand for chemical fertilisers in the USSR remains enormous.

B Soviet chemical fertilisers

Chemical fertilisers are divided into four main groups: a) nitrogenous, b) phosphatic, c) potash and d) complex fertilisers containing all three main elements (and/or some additional substances). We shall first discuss the production of these four main groups in the USSR, while the subsequent section will examine the role of Western technology in sustaining the build-up of the Soviet fertilisers capacity.

In the first group of products, ammonia constitutes the key item because all nitrogenous fertilisers are based on it. In 1975 the Soviet Union produced 12 million tons of ammonia which at that time made it the world's second largest producer. The target for 1980 is 23.1 million tons. This figure, if met, would make the Soviet Union capable of producing about 20 per cent of the present world capacity. As the subsequent section shows, ammonia production build-up in the USSR has been very heavily dependent on the policy of importing Western plants: the importance of this product has been matched by the high priority it has enjoyed with regard to access to Western technology. Some of the ammonia produced in the USSR is applied directly to the land. For example, in 1980 this will be done with about 85,000 tons. The bulk of Soviet ammonia, however, is used for the production of downstream nitrogenous fertilisers. Of these, ammonia nitrate constitutes the largest single type produced in the USSR. Ammonia nitrate is followed by urea and complex fertilisers (containing nitrogen), but as table 16 shows, its share has been gradually declining largely owing to the rising share of the complex fertilisers. In 1975 the USSR produced some 8.5 million tons of nitrogenous fertilisers (in terms of nutrient content),[4] which makes it the largest group (in terms of nutrient). Nitrogenous fertilisers centres are found mostly in the European part of the USSR where the demand is greatest. Since the early 1960s the share of natural gas as the main feedstock has been steadily rising from about 16 per cent in 1960 to nearly 90 per cent at present.

Table 16
Types of Soviet nitrogenous fertilisers

	1965 %	1970 %	1975 %	1980 (plan) %
Total	100.0	100.0	100.0	100.0
Ammonium nitrate	55.3	45.0	44.7	40.1
Urea	17.2	27.3	25.9	27.1
Ammonium sulphate	9.8	7.3	6.8	5.1
Liquid	13.9	14.4	10.9	7.1
Complex	1.3	5.0	10.7	19.3
Others	2.5	1.0	1.0	1.3

Source: *The Chemical Industry of Eastern Europe 1975-80* (Chemical Data Services, 1976).

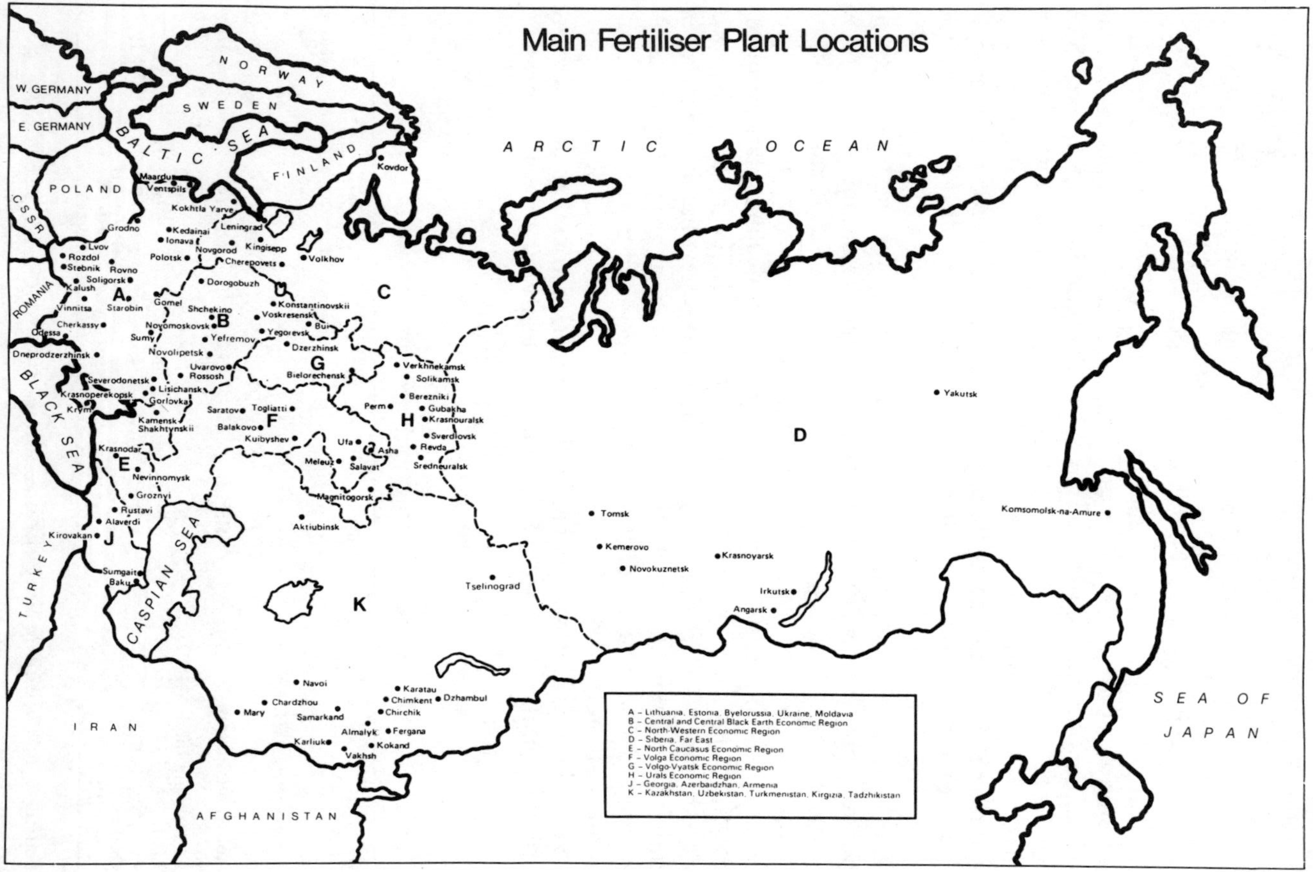

Main Fertiliser Plant Locations

MAIN FERTILISER PLANT LOCATIONS INDEX
(key shown below)

Map A

Cherkassy	AN, U, DAP
Dneprodzerzhinsk	AN, U
Gomel	TSP, MAP, DAP
Gorlovka	AN, U, AS
Grodno	AS, AN, U KNO_3
Ionava	NA, NPK, U
Kalush	K
Kedainai	MAP, SSP
Kokhtla Yarve	NPK, U
Krym	MAP
Lisichansk	U
Lvov	NPK
Maardu	AN, SSP, U
Odessa	The USSR – (see Occidental-deal)
Polotsk	AS
Rovno	AN, NPK, MAP
Rozdol	NPK
Severodonetsk	AN, U, KNO_3, AS
Soligorsk	K
Sumy	SSP, MAP
Starobin	K
Stebnik	K
Ventspils	DAP, (Occidental)
Vinnitsa	NPK, DAP

Map B

Bui	B
Dorogobuzh	AN, NPK
Konstantinovskii	SSP, TSP
Novolipetsk	AN, U
Novomoskovsk	AN, U, NPK
Rossosh	NPK
Shchekino	AN, U, AS
Uvarovo	MAP, TSP
Voskressensk	AN, TSP, MAP, NPK
Yefremov	NPK, DAP
Yegorevsk	P

Map C

Cherepovets	AN, NPK, TSP
Kingisepp	GRP, MAP
Kovdor	P
Leningrad	SSP
Novgorod	AN, U, NPK
Volkhov	TSP

Map D

Angarsk	U
Irkutsk	–
Kemerovo	AN, AS, U
Komsomolsk-na-Amure	U
Krasnoyarsk	AN
Novokuznetsk	AS
Tomsk	AS, AN
Yakutsk	P

Map E

Grozny	AN
Kamensk-Shakhtynskii	AN
Krasnoperekopsk	MAP
Krasnodar	–
Nevinnomysk	AN, U, NPK

Map F

Balakovo	TSP, P
Kuibyshev	P, U, AN, SA, NPK
Meleuz	–
Salavat	AN, U, NPK
Saratov	U, AS, AN
Togliatti	TSP, U, NPK
Ufa	U, NPK

Map G

Bielorechensk	NPK, DAP
Dzerzhinsk	AS, AN, U

Map H

Asha	P
Berezniki	K, NT, U, AN
Gubakha	AS
Krasnouralsk	TSP
Magnitogorsk	AN
Perm	U, AN
Revda	TSP
Solikamsk	K
Sredneuralsk	–
Sverdlovsk	SSP
Verkhnekamsk	K

Map J

Alaverdi	SSP
Baku	AN
Kirovakan	AS, AN, U
Rustavi	U
Sumgait	SSP

Map K

Almalyk	MAP, NPK
Aktiubinsk	SSP, NPK
Chardzhou	SSP, NPK, AP
Chimkent	TSP
Chirchik	AN, U, DAP, AS
Dzhambul	MAP, DFP, TSP, DAP
Fergana	AN, NPK
Karatau	P
Kokand	SSP
Karliuk	K
Mary	AN
Navoi	AN
Samarkand	SSP, MAP, MPK
Tselinograd	MAP
Vakhsh	U

KEY

AN — Ammonium Nitrate
AS — Ammonium Sulphate
DAP — Diammonium Phosphate
DFP — Defluorinated Phosphate
GRP — Ground Phosphate
KNO_3 — Potassium Nitrate
MAP — Monammonium Phosphate
NPK — Complex Fertilisers
SSP — Single Superphosphate
TSP — Triple Superphospha
U — Urea
B — Borate fertilisers
P — Phosphorites (apatite
K — Potash Salts

The use of coal and coke should be totally discontinued by the end of the decade. This development is in accord with the overall aim of 'chemicalisation', namely, to utilise the most efficient sources of feedstock. At the same time it should be noted that the Soviet Union is fortunate in that its natural gas deposits are relatively evenly spread throughout the main area of consumption: see the map 'Main fertiliser plant locations'.

Soviet agronomists maintain that the USSR should increase the use of phosphatic fertilisers in relation to nitrogenous fertilisers and potash. In 1975 the N:P ratio was 1.6:1, while an optimum ratio is thought to be 1:1.[5] A different study concluded that in 1975 the N:P:K ratio was 1:0.7:1 and that it should be changed to 1:0.9:0.7. The planners are responding to these calculations. In the current plan period the rate of growth of nitrogenous fertilisers will be roughly equivalent to that registered during the previous Five Year Plan, the rate of growth of potash will slow down quite dramatically, while the phosphatic fertilisers' output should grow the fastest at an average yearly rate of 14.75 per cent as table 17 illustrates. Soviet phosphatic fertiliser production

Table 17
Average yearly growth in fertiliser production, in %

	1966-70	1971-75	1976-80
Fertilisers, total	12.1	10.2	9.65
Nitrogenous	14.9	9.5	9.65
Phosphatic	9.4	13.35	14.75
Potash	11.7	15.75	6.35

Source: *The Chemical Industry of Eastern Europe 1975-80* (Chemical Data Services, 1976).

consists of processed material such as superphosphate (traditionally the output was dominated by single superphosphate made by sulphuric acid treatment of phosphate rock) or monammonium phosphate (*ammophos*), or of phosphate rock directly applied. Phosphates are produced in granulated, powdered or ammoniated form. The USSR is the world's largest source of phosphate rock. The most important deposits are found on the Kola peninsula where there are large deposits of the world's highest quality apatite. A substantial quantity of this material is exported and about two-thirds of these exports go to the Comecon countries. Another important phosphate rock mining centre is the Karatau basin in Kazakhstan where, paradoxically, a very low grade rock is mined (28 per cent P_2O_5). Other important locations are Kingisepp, Chilisaisk, Yegorevsk, Rudnichnyi, and others. Capacity expansion has been under way in most of these centres and in addition a new major centre will be developed in Yakutia in Siberia, where reputedly there are some of the largest phosphate rock deposits in the world.

Table 18 gives the structure of phosphatic fertilisers production. It reflects the fact that the Soviet Union has been pursuing the progressive policy of expanding the share of technologically most efficient complex fertilisers. The share of straight fertilisers is to decline from 55.7 per cent in 1975 to 36 per cent in 1980. According to Kostandov, the planners are assigning special priority to the increase of production of liquid complex fertilisers, whose share is to rise to 7.7 per cent by 1980 from almost a negligible one in 1975. New technologically advanced products will be introduced, e.g. crystallin (a greenhouse fertiliser), feed ammonium and monsodium phosphates, and the Soviet Union will start manufacturing nitroammophoska by a novel process (i.e novel in the USSR) based on treatment of phosphatic rock by nitric acid.[6] (The traditional method has been that of using sulphuric acid. Hence the imports of the Polish sulphuric acid plants, see chapter 3.)

Table 18

Structure of phosphate fertiliser production, %

	1970	1975	1980 (plan)
Total	100.0	100.0	100.0
Fertilisers	97.8	92.2	90.3
Straight fertilisers	86.6	55.7	36.0
single superphosphate	63.6	39.5	16.0
triple superphosphate	21.3	15.2	19.5
slag	1.7	1.0	0.5
Complex fertilisers	11.2	36.5	54.5
nitroammophoska	1.5	2.4	11.9
ammophos	3.5	26.4	27.2
nitrophoska	1.9	1.6	4.0
nitrophos	3.1	2.8	2.1
nitroammophos	1.1	3.2	1.1
crystalin	–	–	0.3
liquid complex fertilisers	0.1	–	7.7
Feed phosphates	2.2	7.8	9.7

Source: *The Chemical Industry of Eastern Europe 1975-80* (Chemical Data Services, 1976).

Most of the potash fertilisers produced in the USSR are in the form of potassium chloride (the second largest is potassium sulphite). Production is heavily concentrated at two main centres: Soligorsk in Byelorussia, otherwise known as the 'Beloruskali' combine, and Solikamsk-Berezniki combine in the Urals, known as 'Uralkali'. The 'Uralkali' is the largest producer with an output of some 9.5 million tons in 1976. The 'Beloruskali' produced about 8.3 million tons in 1975. Potash is also produced in the Kalush and Stebnik mines in the Ukraine which together add up to about 1 million tons per annum. At all these locations a capacity expansion has been under way. In 1975 the USSR produced about 19 million tons of potash, and if the new planned capacity is on stream by 1980 the country should produce a total of 26 million tons. The Soviet Union is an important exporter of potash: in 1976, for example, it exported 2,500 thousand tons (in terms of nutrient), two-thirds of which went to Comecon.

We have seen that the output of chemical fertilisers, as well as their deliveries to agriculture, have been rising steadily since the launching of 'chemicalisation' in 1958. In the Ninth Five Year Plan (1971-5) alone, the total capacity increment roughly equalled the total pre-1970 capacity of 39.7 million tons. During the current plan period the output is scheduled to rise by a further 59 per cent to a total of 143 million tons in 1980. (It is, however, questionable whether the USSR will be able to fulfil this task. The latest reports at the time of writing indicate that in 1977 the Soviet Union produced 96.7 million tons (in terms of product), which represents a 5 per cent increase over 1976. This amount falls short of the target of 98.5 million tons stipulated in the annual plan for 1977.)

At the same time, the growth of output has been accompanied by a gradual improvement in the composition of production. Some of these qualitative changes have already been touched upon in relation to particular groups of fertilisers, but when discussing the global changes in the output pattern a general comment may be made to the effect that the technologically most progressive complex and concentrated fertilisers with higher nutrient content have registered the fastest growth rates with the result that these more sophisticated types have been steadily improving their share in the total output. It can be seen from table 19 that the proportion of complex and concentrated fertilisers such as nitro-ammophoska, ammophos and nitrophoska constituted 65.7 per cent in 1970, 78.2 per cent in 1975, and will rise to 92 per cent in 1980. Another indication of the improving quality of Soviet chemical fertilisers is the fact that the average nutrient content of fertilisers is to rise from 36 per cent in 1975 to 40 per cent in 1980.[7]

Table 19
Production of concentrated and complex fertilisers
% share in total production

	1970	1975	1980 (plan)
Fertilisers, total	100.0	100.0	100.0
Concentrated, total	61.1	65.2	62.0
nitrogen	29.8	27.9	27.6
phosphate	4.6	4.0	7.2
potash	26.7	33.3	27.2
Complex fertilisers, total	4.6	13.0	30.0
Share of nutrient content in complex fertilisers, %			
nitrogen, N	1.6	4.5	7.8
phosphorus, P_2O_5	1.6	8.1	14.8
potassium, K_2O	1.4	2.3	4.2

Source: *The Chemical Industry of Eastern Europe 1975-80* (Chemical Data Services, 1976).

C Soviet chemical fertilisers and Western technology

This section discusses the role of Western technology in the construction of the Soviet fertilisers industry in general terms, as the information assembled here is based on recent press reports and cannot therefore be completely exhaustive. As in chapter 3, some of the more detailed material is presented in an Appendix (Appendix C).

In the 1960s the Soviet Union purchased several Dutch and Belgian ammonia plants in a more or less sporadic manner. Since the early 1970s Western plants have been installed in accordance with a well-planned programme reflecting in some way the fact that the post-Khrushchev leadership has finally established its attitude towards the agricultural (and agriculture-related) investment. Ammonia plants have, since then, constituted the largest single item on the total Soviet plant purchasing programme. The main feature of the current ammonia capacity build-up is a certain amount of uniformity achieved by heavy reliance on the technology supplied by the Japanese firm Toyo under Kellogg licence. Virtually all of these plants are based on natural gas and their capacity is 450,000 tons per annum (per plant). For details and map see Appendix C. According to Kostandov, these plants, when on stream by the early 1980s, will be able to turn out about 18 million tons of ammonia per annum; this implies that approximately 75-8 per cent of the total capacity of 23.1 million tons will be a direct result of this purchasing programme.[8]

Ammonia also figures prominently in one of the most spectacular cases of East-West economic co-operation. In 1973 the American company Occidental Petroleum successfully negotiated a massive barter deal with the USSR covering a period of 20 years and involving exchange of superphosphoric acid for deliveries of 2.1 million tons of Soviet ammonia per annum as well as large quantities of urea and potash to the United States. Originally it was thought that the project would necessitate the construction of two ammonia pipelines which would move the product from the ammonia plants, situated mainly in Togliatti, to the Soviet ports of Odessa on the Black Sea and Ventspils on the Baltic. Recent developments, however, indicate that in the foreseeable future only one of these pipelines, running from Togliatti to Odessa (2,100 Km long) will be built. It is to be completed during the current plan period and some sections are already in operation. Further equipment currently under construction in the Soviet Union consists of ammonia storage facilities and of a fleet of seven refrigerated tankers. One of these ships has already been launched and at the time of writing first deliveries of Soviet ammonia to the United States are already under way. For a detailed exposition of the deal see Appendix C.

As regards the downstream nitrogenous fertilisers, it may be concluded that the Soviet Union is relatively strong in the area of ammonium nitrate and ammonium sulphate technology. In securing an adequate level of output of urea the industry had to rely on imports of Western plants. The first plants were delivered in the 1960s by Werkspoor of Holland, Mitsui of Japan and Montedison of Italy. The capacity of these plants is 180,000 tons per annum, and they were erected at Salavat, Chirchik, Shchekino, Severodonetsk and Lisichansk. A further six similar plants were ordered in the early part of this decade from Mitsui of Japan and from the Italian firms Technimont and Snamprogetti. Table 20 shows the recent Soviet orders for urea plants. Each of these plants will have a capacity of 495,000 tons per annum.

In the production of straightforward types of phosphatic fertilisers such as single superphosphate, triple superphosphate or nitroammophoska, which until the mid-1970s constituted well over half of all phosphatic fertilisers output in the USSR, the industry relied largely on its own indigenous technology. The extraction of raw material is wholly based on Soviet technology, although this task is not technologically simple owing to large amounts of impurities in the phosphate rock mined; this constitutes a major problem especially at the lesser quality phosphate rock centres such as the Karatau basin.[9] The main reason whv domestic technology must be relied on in the process of

Table 20
Recent urea plant orders

Supplier	Location	Date of completion
Technimont	Gorlovka	1978
	Berezniki	1979
	Novokemerovo (Kemerovo)	1979
Snamprogetti	Togliatti (two plants)	1981
	Novomoskovsk	1980
Toyo	Perm	1980

Source: Reports by *European Chemical News* and *Chemical Age*.

Table 21
Recent orders of phosphoric acid equipment

Firm	Product	Location	Year of completion
Coppee Rust	2 x phosphoric acid plants capacity: 1,000 tp-day each	Cherepovets	1980
Coppee Rust	1 plant – phosphoric acid capacity 1,000 tp-day	Krasnodar	1980
Coppee Rust	1 plant – phosphoric acid capacity: 1,000 tp-day	Meleuz	1980
Spie Batignolles	9 x phosphoric acid concentration units capacity: 150 tp-day P_2O_5 each	Moscow	1979

Source: Reports by *European Chemical News* and *Chemical Age*.

extraction and purification of the rock is that this technology has to be adapted to the specific local conditions and cannot, therefore, be developed abroad.

In the 1960s the Soviet Union bought several phosphoric acid plants from Speichim of France and UCB of Belgium. These plants are located at Voskressensk and Kedainyai. Further purchases included two electrothermal plants for phosphorus from Uhde of the FRG located at Kuibyshev and Chimkent and a triple superphosphate complex from the Dutch firm Niro Atomiser.[10] Recent purchases of phosphoric acid equipment are summarised in table 21.

Potash fertilisers production tends to be based on indigenous technology since it involves relatively simple procedures which, in any case, have to be adapted to local conditions. The machinery the Soviet Union purchased in the 1960s was mostly limited to granulating and mining equipment because this is where the Russians encountered some difficulties. This equipment was delivered by American, Italian, Japanese and German firms and probably installed at Berezniki and the other potash centres. In 1977, however, the Soviet Union ordered a potassium chloride plant for complete treatment from PEC Engineering of France with a capacity of 138 tons per hour to be located at Soligorsk. At the time of writing the USSR was inviting offers for another such plant with a capacity of 2 million tons per annum of potassium chloride to be constructed at Berezniki. Japanese and French firms were bidding for the order. These recent developments may signify a break with the traditional policy in regard to potash technology in the USSR.

In the area of complex and concentrated fertilisers the dependence on Western technology has been relatively high, which would seem to substantiate the proposition that the more technologically progressive the product the greater the dependence on imported equipment. The fertilisers that the USSR produced prior to the launching of 'chemicalisation' were mostly low quality straight fertilisers, and so in updating the technological level of fertilisers output the industry depended heavily on imports of Western plants. However, this is also an area in which the Russians have proved to be relatively fast learners.[11]

In the 1960s the Soviet Union ordered several plants for complex fertilisers from the French firms PEC Engineering and Speichim, the German firm Didierwerke, British Propane Fertilisers and the Belgian firm UCB. These plants are located at Novomoskovsk, Togliatti, Voskressensk, Vinnitsa and Nevinnomysk. Another wave of orders came in 1976. These recent purchases are summarised in table 22.

A precise estimate of the degree of Soviet dependence on imports of Western technology is hard to arrive at because we do not have access to all the information related to indigenous chemical fertilisers plant

Table 22
Recent Soviet orders of complex fertilisers plants

Firm	Product, process, etc.	Location	Date of completion
PEC-Eng and Azote et Produits Chimiques (APC)	Plant for converting super-phosphoric acid into 10:34:0 ammonium polyphosphate liquid fertilisers.	Ventspils	1977
		Rovno	1977
		Vinnitsa	1977
	Capacity: 1,200 tp-day P_2O_5	Cherkassy	1977
		Bielorechensk	1979
	APC technology is used	Voskressensk	1979
	(This order seems to be linked with the USSR: Occidental barter deal)	Yefremov	1979
Speichim	Complex fertiliser plant 17:17:17 type	Rozdol (Novy Rozdol)	1979-80
	Capacity: 800,000 tpa	Meleuz	1979-80
	Process: General des Engrais SA	Bielorechensk	1979-80
		UFA	1979-80
Toyo	NPK 17:17:17 complex fertiliser plant	Dorogobuzh	1979-80
		Rossosh	1979-80
	Capacity: 1,850 tp-day	Rossosh	1979-80
	Norsk Hydro process used	Novgorod	1979-80
		Novgorod*	1979-80
These projects have apparently been undertaken by the USSR alone	NPK fertiliser expansion	Ionava	1976
		Ionava	1980
		Cherepovets	1976
		Novgorod	1976
	Concentrated compound ammonia based fertiliser plant	Fergana	1976-80

*According to some sources this plant will be at Leningrad.

Source: Reports by *European Chemical News* and *Chemical Age.*

construction. Hanson's examination of the various branches of the industry yielded the following rough estimates of the percentages of output based on imported technology: nitrogenous 40; phosphate 5; potash 5; complex fertilisers 65. Some notion of the role of foreign technology may also be gleaned from a statement made by the chairman of 'Tekhmashimport' – the foreign trade organisation responsible for the bulk of chemical equipment purchases – to the effect that in 1973, 52 per cent of Soviet nitrogenous fertiliser production, including 50 per cent of ammonia, 62 per cent of urea and 74 per cent of complex fertilisers, came from the imported machinery. (These figures should be adjusted downwards since about 15 per cent of the imports is thought to have come from Comecon.) The total cost of Western equipment ordered by the USSR from 1960 to mid-1976 has been estimated by Hanson to be at least $2.4 billion and possibly as much as $4 billion.[12] Our findings would seem to confirm the rough estimate of dependence made by Hanson, but in some areas such as ammonia the degree of dependence on Western technology has definitely increased. The same may be said about potash when the recently ordered plants will be on stream.

On the basis of the analysis undertaken in this chapter we may draw the general conclusion that the equipment for the output of the more efficient, and hence more technologically advanced, complex and concentrated types of chemical fertilisers tends to be more import-dependent than the equipment producing less advanced products. This enhances the importance of the role assigned by the USSR to purchases of technology in the West and it indicates the likely future developments in this area of technology transfer for, as we have seen, the output of the more efficient variety of chemical fertilisers is planned to grow at the fastest rate.

Notes

1 M. McCauley, *Khrushchev and the Development of Soviet Agriculture: the Virgin Land Programme 1953-64*, Macmillan, London, 1976, p.119. Another explanation for the relatively slow take-off of fertilisers output could be the fact that fertilisers had to compete with the requirements of conventional munitions production before solid-fuel ballistic missiles were in adequate production (most chemical fertilisers plants were converted to munitions factories during the Second World War).

2 D.W. Carey, 'Soviet Agriculture: Recent Performance and Future Plans' in *Soviet Economy in a New Perspective*, JEC, 1976, p.586.

3 J.P. Cole and F.C. German, *A Geography of the USSR: the Background to a Planned Economy*, Butterworths, London, 1970, p.126.

4 The physical volumes of chemical fertilisers may be expressed either in terms of their actual weight or, more accurately, in terms of the weight of the nutritious substances contained in a particular product. For agricultural purposes it is the latter 'net' weight which really matters because it indicates the amount of nutrients supplied to the crop.

5 *Nitrogen*, no.103, September/October 1976, p.26.

6 Kostandov, loc.cit. (chapter 3, note 5), p.125.

7 Ibid.

8 This figure is based on the assumption that by 1980 the USSR will be producing up to 24 million tons of ammonia per annum which some sources claim to be the case.

9 M.R. Freeman, *The Times*, London, 10 May 1976, p.27.

10 P. Hanson, 'The Impact of Western Technology: a Case Study of the Soviet Mineral Fertiliser Industry' (chapter 2, note 19), table 4.

11 Ibid., p.34. According to a Soviet source, in 1973 about 26 per cent of Soviet complex fertilisers output was based on domestically built machinery. L. Luk'yanov, 'V/O Tekhmashimport dlya sovetskoi khimii', *Vneshnaya Torgovlya*, 1975, no.10.

12 Ibid., p.27 (at early 1970s prices).

5 East-West technology transfer and compensation trading

A Comecon's priorities and the causes of compensation trading

The build-up of the Soviet (and Comecon's) chemical and petrochemical industry is, above all, intended to satisfy the demands of East European industries and consumers. The launching of the programme was motivated by the leaders' awareness of the technological gap between the East and West, but we have seen that in spite of the rapid progress registered since the late 1950s the Soviet Union still lags behind the advanced industrialised countries of the West in terms of the overall composition structure of its output of chemicals. In addition the USSR and the other socialist countries continue to lag behind the West in *per capita* production levels of major groups of products. In 1970, for example, the USSR produced only 6.89 kg of plastics and resins and 0.69 kg of fibres per head of population. The corresponding figures for the major Western producers are given in table 23. With

Table 23
Per capita output of plastics and fibres in the West in 1970

	Plastics, resins (kg per capita)	Synthetic fibres (kg per capita)
USA	42.52	7.93
FRG	70.02	8.35
Japan	49.78	9.93
UK	26.00	6.05

Source: Based on Amann, Appendix, tables 6A.9 and 6A.11 in Amann, Davies, Cooper, *The Technological Levels of Soviet Industry*.

the exception of the GDR, and to some extent Czechoslovakia, the other socialist countries also continue to trail behind the West although they are not quite as far behind as the Soviet Union. Table 24 gives the 1974 output *per capita* data.

Table 24
The 1974 per capita production of plastic and fibres in Comecon, UK and FRG and Comecon projections for 1980

Country	kg per cap Plastics	kg per cap Fibres	1980 projections Plastics	Fibres
USSR	8.37	3.52	21.13	5.0
Poland	11.75	5.82	28.01	10.6
Romania	13.48	7.57	40.00	13.7
GDR	30.12	15.70	56.26	19.4
CSSR	25.46	8.91	53.80	14.4
Hungary	11.05	1.62	27.93	5.4
Bulgaria	15.44	6.78	47.78	13.9
UK	44.05	11.21	–	–
FRG	97.29	15.06	–	–

Source: ICI.

These considerations would lead one to assume that the emergence of Comecon as a major chemicals producing bloc should in no way constitute a threat to the traditional West European and North American producers. In other words, everything produced in East Europe should be consumed there, and moreover, the new capacity in Eastern Europe could in fact create new marketing opportunities within Comecon because, to a large extent, the chemical industry is its own best customer: the build-up of the Eastern petrochemical industries might in fact generate a whole range of new outlets for Western producers provided they respond to the new situation with sufficient flexibility.

The same applies in the area of chemical fertilisers. The case of the USSR is especially pertinent. In 1975-76 the Soviet Union applied 74.1 Kg of fertilisers per hectare, and according to the current plan this amount should reach 114 Kg per Ha in 1980. This compares unfavourably with the advanced Western countries who in 1975-76 applied the following quantities: FRG 235, France 145, Italy 85, Netherlands 305 (Kg/Ha).[1] Moreover, if we remind ourselves that in 1976 only 50 per cent of the grain-sown area in the USSR received any fertilisers at all, it becomes quite clear that the demand of Soviet agriculture is virtually limitless. As in the case of the petrochemicals and their end-products it must be concluded that chemical fertilisers produced in the USSR are planned above all to satisfy the Soviet and Comecon countries' domestic needs.[2] The creation of the agro-industrial complex, of which the fertiliser capacity build-up is an integral part, must be seen in the light of the desire to improve agricultural efficiency, reduce hard currency spendings on imports of grain, and ultimately in terms of efforts to improve the lot of the consumer. The failure of this strategy would have grave political repercussions as the example of Poland in recent years well illustrates (shortages of food, more than of any other group of consumer products, are likely to lead directly to popular disturbances). The socialist leaders are keenly aware of the importance of satisfactory agricultural performance. They also know that any large-scale redirection of resources away from agriculture would sharply contradict their long term economic and political objectives.

Nevertheless, there are at least two reasons why the traditional Western producers are facing major problems in the immediate future. First, the rapid growth of the new industrial bloc is bound to alter the existing *status quo* in international commodity flows. In 1975 Comecon as a whole exported only about 3 per cent of its chemical output and imported about 5 per cent of its consumption. But, granted that the growth of Comecon industries will materialise, it is almost inevitable that this balance will gradually be altered. Within a decade Comecon might export some 5-10 per cent of its output, thus approximating to the levels registered by the other major blocs – Western Europe, North America and Japan. At the same time the structure of Comecon's imports is bound to change as technologically more sophisticated products and intermediates produced on the basis of Western equipment become available, thereby replacing chemical imports from the West.

The second cause of concern arises out of the specific nature of East-West trade in technology. Apart from being desperately in need of high technology, the socialist countries are also chronically short of the hard currency with which this technology may be purchased in the West. This means that the socialist countries are always ready to export commodities which are much needed domestically because these commodities, given the domestic technological deficiencies, are perceived as a potential source of hard currency income and, by the same token, as a source of further inflow of advanced technology. The products manufactured on the basis of these future installations of advanced equipment will, in their turn, stand the best chance of being successfully marketed in the hard currency areas, and the 'vicious circle' of Eastern technological dependence on the West will have completed one of its sequences. These and other similar factors lend the East-West business paradigm very unconventional and unique characteristics. One such feature is the recently fast-spreading practice of compensation, 'buy-back' or 'counter-trade' agreements whereby the hard currency credits, extended by the West in order that the Eastern partner may be able to purchase a particular piece of chemical (or other) equipment, are repaid by the latter in part, or wholly, by the products of the newly commissioned capacity. The socialist countries favour 'buy back' because it obviates the need to provide the repayments of credits in the currency they find so hard to earn given the low competitiveness of their exports. Compensation is, in a sense, a convenient method of significantly cutting short the otherwise difficult and long road towards penetrating the hard currency markets with socialist goods. Given the toughness of competition for socialist orders, intensified by the prevailing climate of economic recession in the West, the companies supplying technology to Comecon find themselves in a position where they find it difficult to resist the East European pressure for 'buy-back'. 'Buy-back' may be said to embody the current extent of East-West systemic convergence: the Western world is suffering from chronic unemployment and the socialist countries from chronic lack of technology; the pressure thus generated on both sides brought about mutual enthusiasm about business opportunities and 'buy-back' is accepted as a price Western industry has to pay if it wants to maintain its ties with Eastern customers.

B The extent of compensation trade

However, the expansion of compensation trade can have very undesirable consequences. First, it represents an essentially alien element within the context of market economies because in practice it amounts to a regression towards bilateralism. It may, therefore, be said that in a sense it helps to 'export' Eastern-style bilateralism into the Western system and thus it completely frustrates any attempts at doing the exact reverse, namely, drawing the East European economies into the Western system of multilateral trade. Secondly, in the paradigm of market economies it can cause serious problems as the 'buy-back' products begin to flood Western markets in ever-increasing quantities, thus contributing to the already substantial over-capacity in the Western chemical and petrochemical industries. In the short run compensation helps to create employment in the Western chemical construction sector but eventually it is bound to hurt Western chemical producers. 'Buy-back' therefore cannot be regarded as a lasting cure of temporary ills; it can, in fact, have very unpredictable and undesirable side-effects especially if this practice is beyond the control of Western governments. But this is precisely what happened in the early part of this decade when Western multinational companies accepted compensation commitments without full realisation of the extent to which it has been practised. At the same time the situation has been aggravated by the unforeseen downturn in business activity from 1974 onwards which substantially reduced the demand for chemicals in the West. Table 25 summarises the data on known compensation agreements signed between the Soviet Union and Major Western companies. (The follow-up of the table giving information on the deals entered into with the other Comecon countries appears in Appendix A.)

Unfortunately it is impossible to produce a precise evaluation of the volume of chemicals which will enter Western markets because not all companies and governmental agencies are willing to disclose all relevant information. However it transpires from the table that the contracts signed between the USSR and Western companies should yield an amount of 'buy-back' material well in excess of $1 billion. To this should be added some $400 million of 'buy-back' coming from the other Comecon countries (see Appendix A) and allowance should be made for undisclosed quantities. Thus the opinion within the industry is that, in all, by the mid-1980s the Western markets could see an annual inflow of East European chemicals amounting to as much as $1.5-1.7 billion at current prices.[3]

A Soviet source stated the value of all compensation contracts signed between the USSR and Western companies to be as much as RbL 20 billion.[4] This seems to be a somewhat inflated figure which probably includes the USSR-Occidental Petroleum long-term contract which, strictly speaking, is not a compensation agreement (essentially it is a barter deal). This figure, moreover, lacks credibility owing to the non-comparability of Western and Soviet statistical categories and, in addition, it is difficult to set an appropriate exchange rate between the US dollar and the rouble. Be that as it may, it has been estimated that the value of contracts in force between 1975 and 1985 – excluding the Tomsk deal with a West German consortium of chemical companies – is somewhat in excess of $4-5 billion. The Tomsk refinery project would add a further $1.0-1.5 billion. The value of contracts – excluding the Occidental 'barter' exchange – therefore ranges between $5 and $6.5 billion. This agrees reasonably well with L. Kostandov's statement, when in England at the start of the Tenth Five Year Plan, that there was a $5 billion 'shopping list' for chemical plant of which $2 billion would be on stream by 1978.

Although this study is concerned with chemicals, it should be pointed out here that these products constitute only one group of products, however important it may be, entering Western markets as a consequence of compensation agreements with Comecon. The value of the deliveries of all products from Comecon could reach as much as $30-35 billion between 1970 and 1985, of which some $25 billion could come from the USSR alone. Over 60 per cent of the Soviet share is represented by exports of natural gas in exchange for large diameter

Table 25
Major compensation deals with the USSR

Company	Year signed	Description of contract	Value	Products to be taken back	Duration of buy back	Marketing by	Annual value of chemical import
Salzgitter, West Germany	1972	Supply of 120,000 ton/year low density polyethylene plant for Kazan	DM126m. ($63m.)	Repayment in low density polyethylene. Between 150-250,000 tons in total	1971 until 1983	Bochako	$ 13.0m.
Salzgitter, West Germany	1973	Supply of 240,000 ton/year low density polyethylene plant for Severodonetsk	DM174m. ($87m.)	Marketing by Bochako of 250-350,000 tons of low density polyethylene in the West. Until 1986	1978	Bochako	$ 25.0m.
Litwin, France	1973	Supply to the USSR of plants to produce 300,000 ton/year styrene, 100,000 ton/year high impact polystyrene and 100,000 ton/year expandable polystyrene for Shevchenko	$120m.	Polystyrene to be bought back over a period of 8.5 years beginning in 1979, at a rate of 24,000 ton/year	1979 to 1987		$ 19.0m.
Occidental, US	1974	Supply to the USSR of 1 million ton/year of superphosphoric acid over a period of 20 years beginning in 1978. Deliveries: 1978 1979 1980 superphosphoric acid 000 ton 10 480 1,000 (also see Appendix C)	$20 billion	In return Occidental will receive 2.1 million ton/year of ammonia for 10 years starting in 1978, about 1.5 million ton/year in the next 10 years and 1 million ton/year of each urea and potash for the entire 20 years. In the first few years the tonnages will be smaller: in 000t 1978 1979 1980 ammonia 350 1,200 2,100 urea 250 250 1,000 potash 380 830 1,000	1978 1997	Interore	$ 241.5m. $ 130.0m. $ 70.0m. $ 441.5m.

Table 25 (cont.)

Company	Year signed	Description of contract	Value	Products to be taken back	Duration of buy back	Marketing by	Annual value of chemical import
Montedison, Italy	from 1973	Supply to the USSR of several plants to be repaid in product. So far the company signed for 11 plants, the original seven were valued at $500m. By now the value is $800m. The plants include: 42,000 ton/year cellulose triacetate at Fergana for completion 1977; 30,000 ton/year polypropylene, Guryev (1977); 270,000 ton/year soldium chloride crystallisation, Sterlitamak (1977); 500,000 ton/year urea, Gorlovka (1978); 150,000 ton/year acrylonitrile, Saratov (1978); 40,000 ton/year chlorofluoromethanes, Volgograd (1979); 40,000 ton/year chlorofluoromethanes, Yavan (1979); 100,000 ton/year polypropylene, Tomsk (1980); 500,000 ton/year urea, Brezeniki (1979); 500,000 ton/year urea, Kemerovo (1979) and a 4,000 ton/year dyestuffs complex at Tambov	$800m.	250,000 ton/year of ammonia, 60,000 ton/year urea, 50,000 ton/year of methanol and 20-30,000 ton/year of acrylonitrile	Urea began arriving in Italy and other countries in 1976 and will continue to 1985. Ammonia started this year for 10 years. Acrylonitrile and methanol will begin in 1980/81 for 10 years to Italy		$ 28.75m. $ 7.80m. $ 5.75m. $ 15.00m. $ 57.30m.
CJB Union Carbide UK	1974	Installation of a 200,000 ton/year high density polyethylene plant at Prikumsk, using UCC technology	over £20m. ($40m.)	Repayment in high density polyethylene from 1980. According to the original agreement UCC was to incorporate the product into its European marketing strategy. Presumably BP will deal with it now. 10,000 ton/year of hdPE	1980-1989		$ 7.0m.
Hoechst Uhde Wacker, West Germany	1974	Installation by Uhde of a 270,000 ton/year VCM plant at Zima, near Irkutsk, based on the Hoechst/Goodrich process	DM100m. ($47m.)	Compensation in VCM from the Kalush, Ukraine, plant which was also built by Uhde and already on stream. Wacker is taking 40-50,000 ton/year VCM from 1976 to 1979 for use in own plants	1976 to 1979		terminates in 1979

Table 25 (cont.)

Company	Year signed	Description of contract	Value	Products to be taken back	Duration of buy back	Marketing by	Annual value of chemical import
Klöckner Huls West Germany	1974	Installation at Zima of a 250,000 ton/year of PVC plant based on Huls know-how	DM150m. ($71m.)	Klöckner markets over 10,000 ton/year of VCM and 8,000 ton/year PVC. PVC already coming in from the Volgograd plant. VCM is not coming yet. Disagree on prices, too high. Klöckner feels will start bringing in VCM in three or four years when market improves. Both will be bought over 10-year period.	1978 to 1991	Klöckner	$ 9.0m.
Creusot Loire France	1974	Installation of two ammonia plants at Odessa and two at Gorlovka, all with 1,360 ton/day capacity	FF1 billion ($213m.)	Compensation with ammonia about 235,000 ton/year. Of this Rhone-Poulenc will take 40,000 ton/year, the rest to be marketed by Intsel	1980 to 1989	Intsel	$ 27.0m.
Marubeni, Japan	1975	Expansion by 25,000 ton/year of the 50,000 ton/year acrylonitrile plant originally built by Marubeni at Novopolotsk (total 75,000 ton/year capacity). Original contract was signed in 1965 and there was no compensation	$10.15m.	Repayment in 10,000 ton/year of acrylonitrile over five years. Deliveries started from Ventspils in December 1977 with a shipment of 5,000 ton	1978 to 1982	Marubeni in Europe	$ 6.0m.
Snia Viscosa Italy	1975	Supply to the USSR of 80,000 ton/year caprolactam plant based on toluene to be located at Chirchik	$180m.	To be repaid in caprolactam at 28,000 ton/year over eight years. Snia will use the product in own plants			$ 28.0m.
Snamprogetti/ Anic, Italy	1975	Supply to the USSR of three urea plants each with a capacity of 1,500 ton/day to be located at Togliatti (two plants) and Novomoskovsk	$200m.	In return Anic will take 100,000 ton/year of ammonia over 10 years, for use in own plants	1979 to 1988		$ 11.5m.

Table 25 (cont.)

Company	Year signed	Description of contract	Value	Products to be taken back	Duration of buy back	Marketing by	Annual value of chemical import
Salzgitter, West Germany	1976	Construction in the USSR (Gorkii) of a 200,000 ton/year ethylene oxide plant using Scientific Design process	DM200m. ($100m.)	A fixed tonnage contract to be repaid in 27,000 ton/year of monoethylene glycol. It is doubtful that all the material will be available because the USSR has only two plants and its needs are higher than this	1979 to 1988	Bochako	$ 10.0m.
Klöckner/Davy Powergas GmbH West Germany	1976	Supply to the USSR (Perm) of a 60,000 ton/year phthalic anhydride plant and a 3,000 ton/year fumaric acid unit. Also catalyst plant	DM700m. ($50m.)	Repayment with 5,000 ton/year of phthalic anhydride, 3,000 ton/year fumaric acid (total amount of plant) and 15,000 ton/year of urea over a period of 10 years.	1980 to 1989	Klöckner	$ 7.0m.
Mitsui, Tokyo Japan	1976	Installation of four ammonia plants each with a capacity of 1,360 ton/day at Grodno, Novomoskovsk, Cherkassy and Rossosh, based on Kellogg technology	$90m.	In return Mitsui is taking back 100,000 ton/year of ammonia and 100,000 ton/year urea both over a 10-year period. Urea began arriving in 1977 and ammonia is to start this year	1977 to 1987	Mitsui trading houses in world markets. Urea is going to S.America, US and Asia	$ 11.5m. $ 13.0m. $ 24.5m.
Rhone-Poulenc, France	1976	Signed a three-part agreement with the USSR: 1. Sale to the USSR of technology and equipment to produce pesticides, fertilisers to the value of FF2.5 billion 2. Receive chemicals from the USSR to the value of FF2.5 billion over a period of 10 years beginning in 1980 3. Sale to the USSR of Rhone-Poulenc products to the value of FF1 billion	Total value FF6 billion ($1.3 billion)	Products agreed on so far are: 40,000 ton/year of ammonia, over 30,000 ton/year of methanol, 10,000 ton/year orthoxylene, 100,000 ton/year naphtha	1981-90 1981-90 1984-93 1979-88		$ 50.0m.

Table 25 (cont.)

Company	Year signed	Description of contract	Value	Products to be taken back	Duration of buy back	Marketing by	Annual value of chemical import
Technip, France	1976	Construction in the USSR (Omsk and Ufa) of two aromatics complexes each producing 125,000 ton/year benzene, 165,000 ton/year orthoxylene and 165,000 ton/year paraxylene	$500m.	Purchase over a period of 10 years of: fixed tonnage orthoxylene 20,000 ton/year paraxylene 20,000 ton/year flexible part benzene 50,000 ton/year diesel oil 290,000 ton/year naphtha 250,000 ton/year The flexible part could now include crude oil also and exclude benzene	1980/81 to 1989/90	Gill and Duffus could be involved in marketing	$ 95.0m.
Krupp-Koppers, West Germany	1976	Supply to the USSR of a 60,000 ton/year dimethylterephthalate plant for Mogilev, Licensor Dynamit Nobel	DM150m.	Dynamit will take DMT, paraxylene, methanol and benzene. It will use paraxylene and methanol and market the other products	1981 to 1990	Dynamit Nobel	$ 10.0m.
Krupp-Koppers, West Germany	1978	Installation of a 120,000 ton/year DMT plant at Mogilev based on Dynamit Nobel process	DM250m. ($125m.)	100 per cent compensation in oil products, paraxylene and DMT		Dynamit Nobel	$ 15.0m.
Uhde/Hoechst, West Germany	1977	Construction of a 35,000 ton/year polyester staple fibre plant at Mogilev. First one in a series of 230,000 ton/year total capacity	DM140m. ($70m.)	Products involved in this deal are DMT, paraxylene and methanol for the time being. Cotton is out (see text)	1981		$ 9.0m.
Klöckner/Davy Powergas GmbH West Germany	1977	Supply to the USSR (Rubezhnoye, Donetsk) of 60,000 ton/year phthalic anhydride plant and 3,000 ton/year maleic anhydride plant	DM100m. ($50m.)	Repayment with 5,000 ton/year of phthalic anhydride, also 3,000 ton/year of maleic (total plant capacity) and 16-17,000 ton/year of urea over 10 years	1980 to 1989	Klöckner	$ 7.0m.

Table 25 (cont.)

Company	Year signed	Description of contract	Value	Products to be taken back	Duration of buy back	Marketing by	Annual value of chemical import
CJB/Union Carbide UK/USA	1977	Construction of a 200,000 ton/year high density polyethylene plant in Kazan based on Union Carbide technology	£50m. ($90m.)	Union Carbide did not give marketing rights for hdPE. Bochako will market ldPE. A total of 240,000 ton from 1983	From 1983	Bochako	$ 16.0m.
Davy Powergas/ ICI/Klöckner, UK/West Germany	1977	Construction in Gubakha and Tomsk of two 2,500 ton/day methanol plants based on ICI technology	over $250m.	ICI and Klöckner will market about 300,000 ton/year of methanol over a period of 10 years. Of the 300,000 ton/year about 200,000 ton/year will be marketed by ICI. In the first two years (1981-82) a minimum of 280,000 ton/year will be shipped in	1981 to 1990	ICI and Klöckner	$ 34.5m.
						TOTAL	$ 922.3m.
UNDER DISCUSSION							
Uhde/Hoechst West Germany		Installation of a 50,000 ton/year polyester fibre plant at Mogilev. Second plant in a series of 230,000 ton/year total capacity	about DM200m. ($100m.)				$ 10.0m.
Tomsk project		Construction at Tomsk, Siberia, of an integrated petrochemical complex	DM2-3 billion ($1-1.5 billion)	Much scaled down version of the original DM8 billion project. It will still be 100 per cent compensation based but in view of the market situation the USSR decided to reduce the project. (See Appendix B)	after 1985 probably until 2000		$ 125.0m.
						TOTAL	$1,057.3m.

NOTES: a) The ECN pointed out that owing to the reluctance of some companies and government agencies the information compiled in the table cannot be regarded as fully exhaustive; b) Market prices which were used to obtain the annual value of imports are as follows: (S/ton) polypropylene 625, polystyrene 800, high density polyethylene 700, low density polyethylene 675, PVC 675, vinyl chloride monomer 330, ethylene glycol 370, dioctyl phthalate 535, acrylonitrile 594, caprolactam 990, ammonia 115, methanol 115, soda ash 122, caustic soda 117.

Source: *European Chemical News*, European Review, 21 July 1978.

steel pipes and other technology related to the extraction and transportation of natural gas. Western interest in Soviet gas deposits in the late 1960s and the contracts signed then could in fact be regarded as the precedent which set the pattern followed in the area of chemical and other trade. About 8 per cent is taken by timber and forestry products and 9 per cent by other groups of commodities. Chemicals represent the second largest group by amounting to about 20 per cent of the total.[5]

C The economic and political implications of compensation trading

The $1.0-1.5 billion worth of East European chemical 'buy-back' products estimated to enter the Western markets annually in the 1980s is not a particularly large amount. It is approximately equal to the present total East European sales of chemicals to Western Europe (just under $1.5 billion in 1977) and just over 40 per cent of Comecon's projected total sales in 1985 or, to put it another way, by 1985 East European sales to Western Europe will amount to an equivalent of one respectable medium-sized petrochemical company.[6] Moreover, the socialist spokesmen like to stress the fact that compensation trade, rather than causing problems in the West, brings definite advantages to the Western firms because they can in this way benefit from stable and relatively cheap supplies of intermediates from Comecon. Despite these claims it is indisputable that 'buy-back' has its negative side in that the Western partners are often forced to import products for which there is little demand from their domestic and foreign customers. The problem has become most acute in the case of products for which there is already substantial over-capacity such as low and high density polyethylene, polystyrene, vinyl chloride, PVC, benzene, styrene, cumene, para- and ortho- xylene, phenol, methanol, DMT, tri- and perchloroethylene, fertilisers, SBR, caustic soda and soda ash. Commodity petrochemicals like polyethylene, polypropylene, PVC, polystyrene or synthetic rubber have little or no intrinsic interest to the major Western producers. The basic difficulty lies in the fact that the days of 15 per cent average growth rate of Western chemical industry seem to be irreversibly a thing of the past. The industry's capacity was planned to meet the growth of West European demand as well as the requirements of export – Western Europe exports significantly more chemicals than all the major industrial areas (the US, Japan, and Comecon countries) and nearly twice as much as North America – and consequently any slowdown of demand hits it particularly hard. In thermoplastics this capacity is at present occupied only to the extent of 55-70 per cent depending on the type of product, and this unfavourable state of affairs has been forecast to last at least until the early 1980s. Moreover, new export-oriented capacity has been (or is being) erected outside Europe in the Middle Eastern oil producing countries (and other developing countries) who wish to increase the value and range of their exports and thus to reduce the preponderance of oil within it.[7] The danger posed by Comecon and the other new producers is that their chemical exports could seriously damage the Western industry's structure and viability by low priced or low cost competition in the same way as has already happened in other industrial sectors, such as the textile industry.

As regards chemical fertilisers, the realisation that the Soviet and East European market is enormous brings little comfort to the traditional producers. Chemical fertilisers tend to be used in regions and countries where they are produced and the world's trade in fertilisers has, therefore, been relatively small. In view of this, to take the most notorious example, the Soviet ammonia which is to be marketed in the West represents too large a volume to be easily absorbed. Recent Soviet ammonia exports have been only about 100,000 tons per annum; they rose sharply in 1976-77 but the forecast increase of up to 2.7-3 million tons by the 1980s is quite staggering. Quite apart from 'buy-back' there are other reasons why the Soviet Union may export its fertilisers in spite of the seemingly limitless demand at home. One of them is that the Russians may not be able to absorb their fertilisers themselves fast enough simply because they may face considerable difficulties in the process of distribution and application to the soil. Soviet sources abound with references to difficulties encountered on this particular front. It should be remembered that the production of fertilisers constitutes only a part of the whole story. The product has to be transported to the farms, stored under prescribed conditions and applied at prescribed times, in the right quantity and form. Since application of fertilisers is a relatively novel activity in the USSR with still insufficiently developed distribution and application channels, it seems reasonable to expect that the surplus will be marketed abroad, especially if there are prospects of substantial hard currency earnings. Past experience has shown that market opportunities have not been unduly wasted by the Soviet Union (and other Comecon countries) even if prices were disadvantageous. The socialist countries are able to manipulate their prices by virtue of their centrally directed foreign trade systems with a relative ease, and the temptation to undercut their competitors on foreign markets is always present.

In spite of the fact that the chemical and petrochemical industry is relatively capital intensive, the threat of further closures of productive capacity (and, hence, threat to employment) that might follow in the wake of increased Comecon exports has triggered off protectionist reactions in the EEC. At the time when the Western companies began to have second thoughts about the virtues of compensation trading with Comecon, the Soviet Union aired its long-term plan of constructing a major petrochemical complex based on a 10 million ton per annum refinery to be constructed at Tomsk in Siberia (see Appendix B and table 25). The complex was to produce 900,000 tons per annum of ethylene and other downstream products and the $4 billion of its total value plus interest rates were to be repaid by 'buy-back' deliveries. These proposals shocked the major West German chemical manufacturers and the German Chemical and Paper Trade Union. Coming at a time when the relatively small quantities of LDPE from the Salzgitter-built plant at Kazan were entering a very weak German market as 'buy-back', the threat of compensation trade on the scale envisaged by the Tomsk project to the health of the West German chemical industry and to employment of workers proved to be too great to be ignored by the West German government. Subsequently the issue was raised during Mr Brezhnev's visit to the Federal Republic in May 1978 and as a result the USSR agreed significantly to scale down the size of the project. It is now valued at $1.0-1.5 billion.

A similar example of such accommodation of interests took place recently in the branch of synthetic fibres where a contract signed between the USSR and a group of German companies (Hoechst, Uhde, Krupp Koppers, Dynamit Nobel and others) in 1976 was subsequently re-negotiated to make it suit the changed requirements of the Western partners. If compensation trading and its rapid expansion during the 1970s came about in consequence of Western weakness under the climate of recession and growing structural unemployment, then the successful pressure exerted by the Western partners in the case of the Tomsk refinery may have turned the tide which up to that point was flowing almost irresistibly in favour of the USSR.

In other countries industrialists and union representatives also began to exert pressure on their governments. In France, for example, the Industry Ministry convened a special meeting in February 1978 which was attended by the leading French producers of chemicals and in the course of which the dangers of compensation trade in sensitive products such as ammonia, polymers and fibres received appropriate attention. This meeting urged the EEC to monitor compensation deals as a prelude to taking action. On the European level, the European Chemical Manufacturers Federation, CEFIC, attempted to channel the demands of the various chemical companies, trade associations and unions into one European voice. In March 1978 CEFIC representatives had their first meeting with the EEC's Industry Commissioner, Vicomte Davignon, during which they raised the question of the implications of compensation trade with Eastern Europe. The various national associations also made their voices heard at the EEC level independently as well as in collusion with CEFIC. Davignon's response has been positive. He recognised the gravity of the dangers faced by the industry and promised the Commission's co-operation with CEFIC on a study of industrial and commercial relations with the states of Eastern Europe. He also expressed his hope that the Commission will be able to open discussions with the Eastern countries concerned and urge them to reconsider their stand on the issue of 'buy-back'.[8]

The compensation trade and its implications are a phenomenon which came about as a result of decisions taken on both sides of the East-West divide. An analysis of this phenomenon, leading to a mutually acceptable solution of the problems that came in its wake, must begin by attempting to understand the positions of both partners. Rather than considering the respective strengths of the bargaining positions of the USSR (and the other Comecon countries) on the one hand and their Western commercial partners on the other, it may be simpler to consider their relations by looking at their respective weaknesses. We have seen in the preceding chapters of this study that the inability of the Soviet-type system to generate satisfactory technological progress constitutes its gravest flaw which, short of fundamental reforms, is virtually incurable. At the same time the imperatives of intensive growth dictate that 'chemicalisation' of the Soviet economy (as an intrinsic part of intensification) should continue unhindered. Its industrial base and R and D capacity are inadequate to sustain the chemical production growth rate of 10.2 per cent planned for the current (Tenth) Five Year Plan, and the USSR invited the Western companies to help it with its tasks. It is too early to judge whether the USSR, with its GNP growth rate slowing down, will be able to expand its chemical industry at the same rate during the Eleventh Five Year Plan as it has done in the past. Be that as it may, there is little doubt that Western technology will be called in to assist further progress, regardless of its speed, and the pressure for 'buy-back' will therefore continue. The USSR would not need to press for 'buy-back' if it could pay for its imported equipment in hard currency but this, of course, cannot be done because the chronic shortage of hard currency is one of the most fundamental features of a Soviet type economy (and this in its turn is a consequence of Soviet technological weakness). And all this is going to persist in the foreseeable future. It has been argued that the USSR has little chance of increasing its hard currency earnings beyond the level of $21-23 billion probably achieved in 1978.[9] The hard currency constraint has been made worse by indebtedness of $15.5 billion and the resulting debt service charge of $3.5 billion in 1977.

Having established that the USSR (and the other socialist countries) will continue to press for a compensation element to be included in its contracts with Western companies, the question now arises whether the latter are in the position to curtail or completely discontinue the practice given the fact that they find it objectionable. We have seen from the example of the Tomsk project that the scale of compensation or the size of planned developments may be modified and scaled down, but it would be completely unrealistic to assume that the suppliers of technology could in the foreseeable future be in a position enabling them to prevent 'buy-back' altogether. The deep-seated economic difficulties which caused the proliferation of 'buy-back' deals in the 1970s show little signs of having receded. Under conditions of intense competition among producers of chemical equipment a company willing to take a greater proportion of 'buy-back' products in return for its services than its rivals automatically increases its bargaining strength (or reduces its weakness), thereby raising its chances of success. Such is the predicament of a bidder operating within the framework of a free market economy when confronted with powerful monopolistic buyers – the Comecon governmental agencies. Even granted that the recession in the West is completely overcome, the attractiveness of doing business with Eastern Europe would be likely to persist because of the promise of access to the natural resources, cheaper labour and vast marketing opportunities offered by the area. The propensity to engage in 'buy-back' is, therefore, as 'systemic' a feature of the Western economies as is the pressure in its favour exerted by the East.

It follows that we can justifiably speak of an emerging situation of economic interdependence roots of which are deeply embedded in the structures of both socialism and capitalism. Although this interdependence is, at first sight, somewhat asymmetrical as East Europe takes a relatively small share of the Western countries' total trade as compared with the West's share in its own trade turnover, it is, nevertheless, a real and ever-deepening relationship which has begun to act as a new and independent factor in the area of East-West relations. The point of crucial importance is that the expansion of East European imports in this decade produced flows of capital, both real and financial. At the beginning of 1977 the USSR and Eastern Europe's hard currency debt stood at $46 billion[10] (and all signs show that it is going substantially to increase in the foreseeable future). The servicing and viability of these debts then call for further and ever-growing Western involvement because the Western creditors (governments and private banks) as well as the manufacturers of technology have a proportionately increasing stake in the development of the Eastern economies; a collapse of the growth strategy in the East would have unwelcome repercussions in the West.

Under such conditions the main task is that of attempting to find mutually acceptable means of overcoming the present difficulties. One possibility is to aim at a greater degree of complementarity by letting the Western industries specialise in the production of technologically more sophisticated products while the Comecon countries would deliver the basic intermediates and/or raw materials. Complementarity has already become the key concept in the EEC-Middle Eastern negotiations[11] and it has already been proposed that a meaningful reciprocity in the realm of East-West commercial relations could be achieved by asking the Comecon countries to refrain from investing in the production of a particular product and thus to leave room for Western imports.[12] Should this approach be successfully applied in some areas of East-West commerce then there would be no reason why it could not be extended into the chemical trade too. The international co-ordination of investment plans is by no means an easily attainable objective, and since we are talking about co-ordination between two ideologically and politically hostile blocs the task becomes twice as difficult. Nevertheless, at least a greater degree of openness in the process of drafting the East European plans, in order that Western firms may assess Comecon markets and shape their own investment policy accordingly, would be a welcome start to such co-operation and would, in any case, comply with the commitments enshrined in the 'Basket Two' of the Final Act of the Helsinki Agreement signed by all CMEA countries in August 1975.[13] In such a discourse the socialist countries would almost certainly oppose any proposals made by the West which in their eyes would seem to imply that the West wishes to prevent them from significantly upgrading the technological level of their chemical output. The objectives of 'chemicalisation', after all, are precisely the opposite – to close the technological gap between the East and West – and this study has already discussed the basic fact of life, namely, that no country is forever prepared to swallow its pride and indefinitely to occupy the position of a technological follower.

On the other hand, the way towards closing the gap leads through an ever increasing Eastern dependence on the services made available by Western companies and this is an equally weighty fact of life. Some accommodation of interests, therefore, has to be achieved. The socialist countries have a vital stake in continued good relations with the West and do not wish to pursue their own interests (i.e. pressing for 'buy-back' at any cost) too hard as this could produce an undesirable backlash and hardening of attitudes in the West, thus threatening the very basis of East-West commercial *détente*. It is significant that the most trade-dependent East European member of Comecon, Hungary, has already declared that it is opposed to an indiscriminate use of compensation as a tool for generating hard currency revenue, and the

Hungarian Ministry of Foreign Trade has published what may be taken to be its 'code of conduct':[14]

> 1 Those forms of countertrade transactions – linkage of exports and imports – should be eliminated, where such products are imposed on the partners, which the enterprise itself is unable to sell, or where it has no marketing organisation for its proper distribution.
>
> 2 Countertrade deals are considered useful, if the foreign supplier is the user of the counteritem offered, or if it has an established commercial organisation for its distribution.
>
> 3 Uneconomical, obsolete, not competitive products of inferior quality should not be offered as counteritems, because it is not desirable to conserve outdated production through countertrade.
>
> 4 In the case of purchase of capital goods, the offer as counteritems of goods produced with the equipment bought should serve the purposes of improvement of quality, guarantee or the introduction of the goods on the market.
>
> 5 When entering into countertrade transactions, enterprises should rely to a greater extent on the assistance of the Hungarian trade representation in the country concerned.

We should be wrong to assume that the Hungarian position will automatically be adopted by the rest of Comecon and notably by the most important compensation trader, the Soviet Union. Nevertheless there are signs that the USSR too appreciates fully the reluctance of Western companies to take on 'buy-back' commitments in some of the particularly sensitive products; the USSR no doubt wishes also to preserve its reputation as a 'good trader'. While East-West investment co-ordination may be a thing of the distant future, the scaling down and modification of projects and/or spreading of 'buy-back' quantities to be taken by the West over increasingly longer periods of time are probably the most suitable tools readily available today.

Notes

1 Eurostat., *Yearbook of Agricultural Statistics*, 1977 (Brussels), p.69.

2 In the past the percentage of Soviet fertilisers going to export has been only about 8-9. Most of it went to the CMEA or LDCs such as India, Pakistan and Turkey.

3 *European Chemical News*, European Review, 21 July 1978, p.32. The upper limit of this estimate was quoted by the ICI deputy chairman, John Harvey-Jones, in *Financial Times*, 6 November 1978.

4 *EcoTass*, November 1977, no.45, p.6.

5 Derived from a draft of a forthcoming OECD report on 'Countertrade Practices in East-West Economic Relations'.

6 *European Chemical News*, European Review, 21 July 1978, p.32.

7 L. Turner and J. Bedore, 'Saudi and Iranian Petrochemicals and Oil Refining: Trade Warfare in the 1980s?', *International Affairs*, vol.53, no.3, June 1978, pp 572-86.

8 We are indebted to P. Marsh for information concerning the protectionist pressure on the EEC level. See his 'Multinational Corporations and the development of East-West Detente: a case study of corporate involvement in Western foreign economic policy towards the Soviet Union and Eastern Europe', an unpublished PhD thesis, pp 253-6 (University of Manchester, Department of Government, 1979).

9 CIA, *USSR: Hard Currency Trade and Payments, 1977-78*, March 1977, p.10.

10 R. Portes, 'East Europe's Debt to the West: Interdependence is a Two-Way Street', *Foreign Affairs*, no.4, July 1977, p.752.

11 Turner and Bedore, loc.cit. (above, note 6), p.583.

12 J. Pinder, 'Community and Comecon: What Could Negotiations Achieve?', *The World Today*, May 1977, pp 176-85.

13 M. Kaser, 'Monitoring East-West Trade for Madrid' in *Contemporary Review*, May 1979, pp 225-30.

14 The Hungarian Ministry of Foreign Trade, Budapest, November 1977, Department for Exchange, Financial and Price Matters, GATT document L/4633.

CONCLUSIONS

6 Implications of East-West transfer of technology

In this final chapter we shall take up some additional points not previously discussed in this study which have emerged as a consequence of recent debate and scientific research into the various facets of East-West relations and technology transfer. It is hoped that the following discussion will also provide a suitable conclusion to our illustration of the East-West technology flow within the Soviet 'chemicalisation' programme.

A The measurement of Soviet dependence on Western technology

In chapter 2 of this study we have referred to Sutton's findings of the Soviet Union's past record in developing new technological processes. In the same work Sutton argues that Soviet technological dependence on the West is total and in general terms he implies that any strictly centrally planned Soviet-type economy would choke of technical inertia were it not for the willingness of market-based economies to supply it with technical services and productive units.[1] Indeed, the evidence on which such a conclusion is based is sufficiently weighty and convincing and, moreover, we have been able to identify a whole host of features peculiar to strictly centrally planned systems which impede technological innovation and its diffusion. In so far as such features may be regarded as systemic, that is to say their elimination would involve a fundamental reform of the system, the centrally planned economies may be said to be 'structurally' dependent on the supplies of technology produced within systems not regulated strictly by central planning.

However, it is conceivable that the anecdotal evidence amassed by Sutton could to some extent be modified. Amann, for example, argues that this could be done at least on one occasion and he sees as a major drawback the fact that Sutton's work is based predominantly on Western-published material.[2] At any rate, it would probably be possible to modify the interpretation of his evidence and thus the tenor of the conclusions he draws. It would be untenable seriously to argue that *everything* the USSR has achieved is a *direct* result of imported, or otherwise acquired, Western technology. Clearly the existing technological base which the Soviet Union possesses must be the result of both indigenous and foreign inputs. But if this is the case then we should be wide of the mark if we regarded the Soviet industrial potential either (a) as a result of strictly Soviet efforts, or (b) as a direct result of Western technology acquisitions. The Soviet representatives tend to adhere to the former proposition, while Sutton might conceivably (for the sake of the present argument) be accused of over-emphasising the latter position. Whether or not the above statement is justifiable is hard to establish because of the Soviet obsession with secrecy and unwillingness to talk about the USSR's dependence on Western technology. Even when the subject of importation of this technology is openly discussed, the Soviet officials as a rule stress the relatively minimal size of Western imports in proportion to the size of total inputs or in proportion to the magnitude of their economy. Thus the USSR deputy minister of foreign trade claims that the total volume of Soviet imports from advanced capitalist countries currently amounts to less than 1.5 per cent of the USSR's gross product.[3] Indeed it is important to realise that the volume of this trade is dwarfed by the size of the Soviet economy and by the volume of its output and inputs. Hanson estimated the share of imported Western machinery expenditure in the total Soviet machinery investment.[4] His estimates are reproduced in table 26. It is apparent from the table that the contribution made by Western imports, although steadily increasing, is very small.

On the other hand, it is essential that a distinction is made between the approach which considers Western imports in terms of overall inputs and an approach which treats Western imports in the light of benefits expected to be derived from them. Given the chronic scarcity of hard currency it is a commonsense observation that the Russians must exercise great caution in deciding which items to purchase in the West, which implies that machinery imported from the hard currency markets is expected to perform much better than the comparable Soviet or East European type, or indeed, that such machinery is simply unavailable within Comecon. Trade statistics by themselves cannot give an accurate picture of the benefits the Soviet economy gains because they do not capture the fact that the productivity of the imported machinery is much higher than that of its domestic (or Comecon's) counterparts.

Research based on the use of econometric models of the Soviet economy has been pursued in the West in recent years with the aim of throwing more light upon this problem. Perhaps the best known such study is that undertaken by Green and Levine.[5] Their methodology involved the calculation of the stock of imported and domestic machinery separately and the statistical estimation of the relative

Table 26
Soviet imports of Western machinery and transport equipment 1955-76

	Value of imports of Western machinery in US dollars, FOB, current prices (millions) (1)	Estimated rouble value, in 1969 investment prices (millions) (2)	Imports as share of domestic machinery investment in the following year (per cent) (3)
1955	104	148	2.0
1956	139	188	2.4
1957	128	166	1.8
1958	123	158	1.6
1959	177	227	2.2
1960	310	393	3.4
1961	390	472	3.6
1962	436	510	3.6
1963	402	467	2.8
1964	489	561	3.2
1965	366	421	2.2
1966	395	436	2.2
1967	457	499	2.4
1968	639	721	3.2
1969	889	966	3.8
1970	905	913	3.4
1971	840	796	2.8
1972	1,126	959	3.0
1973	1,574	1,091	3.2
1974	2,094	1,316	3.4
1975	4,184	2,287	5.6
1976	4,259	2,282	5.5

Notes: a) The figures in column (1) are based on Western trade returns. They represent imports covered by the Standard International Trade Classification, section 7. Finland is excluded owing to its special barter relationship with the USSR.
b) The methodology used in calculating columns 2 and 3 is explained in P. Hanson, 'The Import of Western Technology' in A. Brown and M. Kaser (eds), *The Soviet Union Since the Fall of Khrushchev*, 2nd ed., Macmillan, 1978, p.31.
c) The 1976 figure in column 3 relates to planned rather than reported investment.

Sources: Philip Hanson, 'International Technology Transfer from the West to the USSR' in *Soviet Economy in a New Perspective* (JEC, 1976), p.796. Philip Hanson, 'Western Technology in the Soviet Economy', *Problems of Communism*, November/December 1978, p.22.

contribution of these separate components to the growth of Soviet industry. In their initial study the authors concluded that Western machinery is approximately 15 times more productive at the margin than indigenous machinery and moreover that the growth of industrial production from 1968-73 would have been only 28.4 per cent without those additional imports of Western machinery (instead of the actual

33.4 per cent) or, in other words, that approximately 15 per cent of the growth rate would have been foregone. These calculations were subsequently revised but even then it was shown that Western machinery installed in Soviet industry is about 8 times more productive at the margin, which implies that about 2.5 percentage points of the 1968-73 industrial output growth rate would have to be subtracted.[6] Admittedly these studies rest on a number of debatable assumptions, but they do establish beyond doubt that the transfer of technology is much more important to the Soviet economy than a superficial examination of trade statistics would suggest.

Although this approach, and the calculations supporting it, prove that Soviet investment in Western technology results in relatively high returns, there are several factors which may reduce somewhat the net economic effect of importation of Western technology. Perhaps the main such factor is that the installation and successful commissioning of foreign technology in the USSR requires additional expenditure on the necessary infrastructure, and frequently a far-reaching expansion of related industrial output. Moreover, even the process of importing machines in order to produce machines at home must be accompanied by considerable R and D expenditures. In the absence of technology imports this additional investment could be allocated to domestic innovation and diffusion. The results might substantially offset the losses from the absence of technology transfer.[7] This argument may be taken further. There is no doubt that in the long run the injection of Western technology must have a positive, stimulating, effect on the overall Soviet economic performance but in the short term, in the course of installing foreign machinery, the large Western-assisted projects such as the Volga Automobile Plant or the Kama River Truck Plant might compete for scarce resources with the other high priority sectors (namely defence) if the resource-demanding function of such projects exceeds their resource-releasing function. At least one observer has suggested that this may indeed be the case.[8] Moreover, projects such as those related to the expansion of the automotive industry in the Soviet Union usually tie up more resources over a period of time as the need for better services, roads, and auxiliary motoring industries arises. The construction of the Kama River Truck Plant, for example, required over $1 billion in hard currency expenditure and $2-3 billion in domestic expenditures, in other words a ratio of foreign to domestic expenditures of 1:2 or 1:3.[9] At the Volga automobile plant this ratio was about 1:2.[10] At the same time such projects employ the highly qualified manufacturing engineers and project managers who are relatively scarce in the USSR, while the relatively plentiful Soviet R and D scientists and engineers are only incidentally concerned with them. If the automotive industry were expanded in the future these initial domestic costs as well as foreign technology expenditure would soon be dwarfed by the additional investment needed in order to improve the automotive infrastructure. It is interesting to note that the current Five Year Plan in fact does not foresee the growth of the automotive industry, especially of the passenger cars industry, on the scale set by the precedent of the previous plan period. It may be that one of the reasons for this is that the planners are aware of the ever-growing needs of these industries and the leadership decided to check further expansion in this direction lest it should compete with the high priority military sector. In any event such a development would contradict the anti-consumer bias of the system.

The other factor reducing the net benefit derived from importation of Western equipment is that the USSR is unlikely to utilise the imported machinery efficiently.[11] In addition the system is hardly able to diffuse this technology at a satisfactory rate, basically for the reasons which we have already outlined when discussing the obstacles to promotion of innovation in general. In the final analysis it makes little difference whether an innovation is of domestic or foreign origin. Initially the large projects such as those related to the automotive industry and requiring large outlays of domestic and hard currency investment must be treated as high priority schemes and are likely to perform at a satisfactory level of efficiency. For the same reason their marginal productivity is likely to be very high. A growing number of Western-assisted projects and an increased share of Western machinery might be associated with a less striking disparity between the marginal productivity of imported and domestically produced equipment. Indeed the Western import share is so small that this dispartiy might fall very sharply as the share was raised.[12]

The chemical and petrochemical, as well as petroleum products, industries have been the major claimants of the Soviet imports of Western machinery. Green and Levine's study, quoted above, singled out these two industrial sectors as two of the three main groups of Soviet industries to which the equipment purchased in the West was allocated (the third main group is the machine-building and metal-working industrial sector). It appears that in most of the years between 1960 and 1974 the value of estimated stocks of Western equipment in these two industries combined accounted for over half of the total value of Western machinery installed in Soviet industry as a whole. (This proportion was much lower, about one-third of the total, in the first four years of the period but during that time these groups registered the highest growth rates of the stock of Western machinery.) This clearly illustrates the importance attached to 'chemicalisation' in relation to the overall strategy of industrial development in the USSR and it also gives us a rough idea of the role performed by Western chemical and petroleum equipment in enhancing Soviet industrial growth.

As regards the magnitude of the role of Western equipment within the chemical industry alone we may consider the following data. Between 1966 and 1975 the USSR imported RbL 3,415 million worth of chemical equipment of which RbL 1,954 million, or some 57 per cent, came from the industrialised countries of the West and the remainder from Comecon.[13] This proportion may be taken to be representative of the pattern prevailing since the launching of 'chemicalisation', although in individual years it may have varied. In 1970, for example, the share of Western imports was as 'low' as 38 per cent while in 1976 it reached a staggering 73 per cent.[14] Moreover it appears that the value of the imported equipment (both Western and Eastern) has on average constituted about 34 per cent of the total rouble value of equipment produced domestically and imported between 1958 and 1974[15] (which implies a foreign to domestic equipment ratio of about 1:2). This means that approximately 20 per cent of the total rouble value of the Soviet chemical industry's equipment has been imported from the West.

B Technology transfer and East-West relations

The rate of technological change and the rate of diffusion of technical innovation are today recognised as a major source of economic growth. However, it seems that international transfer of technology, which is, in a sense, a process of diffusion of technical change within the world economy as a whole, has not as yet won adequate recognition as an important source of growth.[16] In chapter 2 (section B) of this study, we have briefly touched upon some of the main implications of Gomulka's approach to the East-West technological gap. A further implication of this approach is that the international transfer of technology is given full recognition of its status among the other important factors on which economic growth depends, such as capital formation, investment, indigenous R and D and economic industrial management in general. Subject to certain conditions – above all a satisfactory degree of openness to foreign technology and preparedness for its infusion – the transfer of technology from a developed country to a medium developed country becomes the dominant source of growth in the medium developed country. The share of technologically advanced imported machinery in total domestic investment may thus become a determinant of growth: an increase of this share should raise the growth and vice versa.

This upgrading of the importance of international transfer of technology within the realm of economic theory should be accompanied by a parallel movement within the paradigm of international, and in our case East-West, relations. The remaining discussion will consider the present state of East-West relations in the light of this increased emphasis on the transfer of technology.

The net effect of the policy of importing Western technology may not be too great in the case of the USSR because of the size of the Soviet economy and the relatively small share of imported machinery in total investment, but it is much greater in regard to the smaller and more trade-dependent members of the CMEA. These countries are therefore interested even more than the USSR in maintaining and expanding their trade with the Western developed nations. In addition there are important political differences between the Soviet Union and the other socialist countries of East Europe. The demonstration effect of the Western standard of living is much more powerful in the socialist countries of East Europe than in the USSR and, as recent events in Poland have shown, the East European governments cannot afford to risk the lowering of consumption levels because their political control over the population is not on the whole as effective and complete as in the USSR. The Soviet Union needs Western technology in order to implement its large-scale industrial campaigns, such as 'chemicalisation', or for the purposes of extraction, processing and transportation of its raw materials and fuels. It possesses large deposits of these materials and they constitute the most important group of its exports both to Comecon and to the West.[17] The other East European countries on the whole are not endowed with similar natural resources[18] and their primary aim in pursuing the policy of importing Western technology is to increase their exports of manufactures to the hard currency Western markets. They are, therefore, strongly motivated to produce consumer goods of adequate quality both for export to the West and for the domestic and CMEA markets. This has been reflected in the more open policy towards Western Europe pursued by some of the Eastern countries (notably Hungary, Poland and Romania).

Western observers of the CMEA scene usually suggest that the policy of importing Western technology is a more palatable alternative to the other method of promoting technological progress (and hence intensification), namely that of implementing the fundamental reforms of the economy required in order to improve the rate of domestic innovation.[19] It may, however, be suggested that even such fundamental reforms would not immediately succeed in terminating Eastern dependence on Western technology. In any case such reforms would not preclude technology transfer; on the contrary, it is a part and parcel of such reform that the isolation of the economy from external influences is as far as possible curtailed. It can, in fact, be maintained, as some writers do,[20] that the Soviet leadership is pursuing both strategies simultaneously: the import of large quantities of foreign technology is accompanied by piecemeal reform of the existing organisational structure of Soviet industry. Both alternatives are thus by no means mutually exclusive but, on the contrary, one complements the other. Nevertheless, it is the relative importance of one over the other that counts, and in this respect it must be concluded that at least in the USSR import of technology clearly receives an immeasurably higher priority than the promotion of indigenous technological innovation.

In chapter 3 we posed the question of whether the Soviet Union and the other CMEA countries could revert into the isolation characteristic of the years of Stalin's rule. Our examination of the role of Western technology in sustaining 'chemicalisation' seems to suggest that the Soviet Union, and probably even more so the other socialist countries, will not and cannot return to isolation because their technological dependence on the outside world would strongly militate against such a trend. The nature of the equipment supplied to the USSR is such that it cannot be scaled up or replicated without profound understanding of the processes involved and without possessing an industrial base capable of turning out the necessary high quality components.[21] However, this applies not only to the qualitative side of the question but also to its quantitative aspect. The scale of technology transfer in the chemical industry alone is such that its termination would considerably harm the further development of the socialist economies. To sum up, our examination of one facet of the socialist intensification supports the view that the current East-West co-operation will continue as long as intensification itself figures high on the priority list of the East European economic planners. The policies of intensification, however, cannot be discontinued because the East European governments are deeply committed to closing, or at least substantially reducing, the East-West technological gap. Although there are important differences between the USSR and the remaining European members of Comecon in regard to the objectives of their policy of purchasing advanced technology in the West, and the degree of the economic and political importance of such transfer may vary from country to country, it may be concluded that all the socialist countries have one thing in common: their economic development would considerably suffer both in the short and long terms should the scope of industrial co-operation with the West be substantially reduced. This would, in addition, bring about the risk of increased political instability in the region, a risk the socialist leaders cannot afford to take. Thus perhaps the most important conclusion from our exercise of observing East-West relations through the prism of the international transfer of technology is that the current process of East-West *détente* is here to stay. Occasional periods of increased tension will certainly occur but the force of Eastern attraction to the sources of Western technology is far too great to be overcome by temporary outbreaks of the cold, or 'lukewarm', war. Moreover, we have argued in chapter 5 that the East European indebtedness and 'buy-back' commitments have already led to a substantial degree of interdependence which is likely to increase rather than diminish regardless of what the Comecon leaders might wish. If East-West technology transfer is the main pillar of *détente*, then this process will almost certainly deepen and further accelerate rather than slow down or reverse altogether.

The transition from cold war to *détente*, from economic warfare to interdependence, has been accompanied by a profound change of attitudes on both sides of the East-West divide. As Wilzcynski has shown,[22] it is above all the large multinational companies and big banks, institutionally and financially supported by the various national governments, which have become the main channels of the Eastward flow of capital and technology. Thus the traditional foes and ideologically diametrically opposed entities – the socialist states and capitalist multinationals – have become the main engines of change in the realm of East-West relations. If the end of the cold war was a consequence of the technological strength of both blocs in the sphere of military technology, then peaceful coexistence and *détente* have been the result of technological weaknesses and economic failures. It is no coincidence that the recent expansion of East-West trade in technology has occurred at a time of economic recession in the West. The thesis expounded in this study is that the East-West flow of technology is primarily due to the failure of the socialist systems to generate a more satisfactory rate of indigenous innovation, but we also maintain that the Western side responded in the way it did because, among other reasons, the temporarily ailing Western economies have been badly in need of investment outlets and new markets.

Given these profound changes in East-West relations it might be useful briefly to assess their importance in the light of the notion of East-West convergence. The idea of convergence, briefly speaking, postulates that in the course of the scientific and technological revolution, and owing to the process of industrial development generally, the capitalist and socialist worlds are propelled along ultimately and inevitably converging paths to a point where there will be no significant institutional, socio-economic and, ultimately, political differences between the two systems. There will be, the theory maintains, a kind of homogeneous industrial society directed by an institutional set-up which will in some respects correspond to the present-day socialist systems while in other respects it will possess the features of contemporary capitalist societies. Despite, or perhaps because of this, the new system will be a distinct and unique entity. Although the idea of convergence is as old as the USSR itself,[23] it became especially fashionable in the early 1960s when for a time it seemed that the socialist countries were on the verge of setting off an irreversible process of market-based reforms while the Western governments played an increasingly greater role in managing their capitalist economies. Subsequent reversals of such policies in the East, culminating in the termination of the Czechoslovak reform movement in 1968, as well as the persistent Soviet claims to the effect that East-West *détente* in no way implies an abandonment of the class struggle on the international (as well as internal) arena, have discredited the convergence thesis to such an extent that it is now necessary to look afresh at the East-West scene with the aim of filling the vacuum created by its collapse.

On the basis of our examination of the role of technology transfer in the realm of East-West relations we may postulate the existence of two distinct, but interrelated, facets of convergence. It may be said that positive steps in the direction of convergence have been taken because in importing advanced technology the socialist countries have been aiming at accomplishing such changes in the combination of their production factors as would make their economies more like those of the industrially advanced countries of the West. These efforts have been figuring high on the priority lists of the CMEA governments, but this economic facet of convergence has not so far been matched by systemic, and by implication political, convergence because such a move would, from the point of view of the present-day East European leaderships, require the taking of unacceptable risks in the realm of politics. The socialist countries have, of course, found it necessary to render their systems more responsive to the needs of the intensive strategy of growth, but there are very definite limits as to how far such systemic changes could go without the loss of political control on the part of the governing elites. The interrelationship between the economic and political facets of East-West convergence is very complex and a careful, country by country, examination of its nature would probably show that it varies from case to case and changes at different periods of time (in the same way as does the relationship between politics and economics in general). Such an examination would take us well beyond the bounds of this study, but for the purposes of the present analysis it will suffice to say that in dealing with Western companies and governments the Soviet and the other East European governments must continually balance the costs and benefits of such co-operation, and the evidence shows that so far they must have concluded that the course of economic convergence can successfully be pursued without the necessity of fundamentally reforming their economic and political systems. Moreover, it could be argued that, at least at the present time, the socialist countries import technology in order to avoid the implementation of such reforms (although, as we have pointed out above, co-operation with the West and systemic reforms are by no means mutually exclusive strategies; in fact, in the long run they are complementary).

Be that as it may, our assessment of recent developments in East-West relations must accommodate both of the facets described above. On the one hand it has to take account of the movement in the direction of what we have referred to as economic convergence while, on the other hand, it has to reconcile such a movement with an equally powerful pressure exerted in the direction of continued ideological, systemic and political divergence. Perhaps the ideal way to accommodate these two factors is to consider the whole situation in terms of a rapidly developing *convergence of interests*. Such a formula implies that while fundamental systemic and ideological differences between East and West persist and, to be realistic, are unlikely to diminish in the foreseeable future, there is nothing to prevent meaningful co-operation in the field of industrial development. The agreement to disagree has been one of the fundamental principles of political and commercial practice in the West and it seems that it has now been applied in the area of East-West relations, thus enabling both sides to engage in what has aptly been described as 'transideological co-operation.[24]

In our Introduction we have proposed that this study should approach the subject of East-West transfer of technology in a way that interprets it in terms of a flow of highly valuable resources from the market-based Western systems to the centrally planned Soviet and other East European systems. In chapter 5 we have suggested that the more recent developments in the field of East-West industrial co-operation have produced a situation of interdependence between East and West and we have argued that the factors that have brought about such interdependence are attributable to the systemic features of *both* the market and the centrally planned economies. The notion of convergence of interests introduced at this stage plays a key role within this overall theoretical framework, for if there is such a collusion of interests then, in reference to our approach, it is, above all, a convergence of Eastern and Western sides' interests in alleviating the deficiencies deeply ingrained in their respective market-based and centrally planned systems of economic direction. To take this argument to its conclusion, the Eastward movement of high technology and the Westward flow of 'buy back' commodities have raised a wide range of problems effective solution of which requires a continuous East-West dialogue and a willingness on each side to appreciate the other partner's difficulties. The East-West technology transfer, therefore, has led to the evolution of a *system* of contacts and consultations necessary for sustaining and managing the ever-growing 'transideological' industrial co-operation. If, for this reason, the East-West transfer of technology is visualised not as a process but as a pivotal component of a system designed in response to the need to alleviate the shortcomings of both capitalism and socialism, then both market-based and centrally planned systems are reduced to the status of sub-systems generating the necessary conditions under which the system of East-West industrial co-operation was able to spring into existence and subsequently continue to live a life of its own. It is to be hoped that this genealogically higher system of East-West economic co-operation will exert greater and greater influence on the attitudes and policies shaped within its constituent parts and that the economic *détente* will continue to fortify the *détente* in the political and military spheres.

Unfortunately the perpetuated state of ideological and political divergence cannot but militate against the current Western policy of letting the companies involved in East-West transfer of technology supply the socialist countries with the high technology they themselves are not able to develop. Undoubtedly powerful arguments against this technology transfer may be constructed on economic, political and moral grounds.[25] Indeed it seems incongruous that the Western countries should help to support political and economic systems which in virtually all conceivable respects are alien and inimical to their own, and it is irrational that such a policy is accompanied by enormous defence expenditures whose main purpose is to shield the West from the military threat of the Warsaw Pact countries.[26]

However, on balance the arguments in favour of East-West technology transfer are probably weightier. First, it would be unrealistic to assume that the USSR and the other socialist countries could be pressurised into making substantial external political concessions or internal systemic changes by the Western denial of technology supplies. In this study we have argued that because of the persisting (and persistent) 'lop-sided' nature of the East European systems it is impossible objectively to assess the degree to which their apparent stability glosses over the latent forces of their inherent instability. Under these circumstances it would be hazardous to exert such a pressure because the final outcome of its use would be very unpredictable and well beyond the control of those who would have initiated it. The probability that the external pressure would play into the hands of the more hard-line elements within the East European elites, thereby slowing down or reversing the process of *détente*, is too high to make the complete termination of technological co-operation with the East a viable proposition. Secondly, any effective measure applied in the West in order completely to stop the Eastward flow of technology would in practice imply virtual abolition of the basic principles of the free market economies (as anything less drastic would simply not work and, as in the past, the technological embargo would almost certainly be gradually eroded). Thus in the final analysis the implementation of such measures would mean that it would be the West which would become more like the socialist countries and not the other way round.

This, however, is not the same as to suggest that the position of the Western countries leaves them with no leverage at all or that the practice of providing the East European countries with technology should not be subjected to continuous profound questioning and revision. On the contrary, in the absence of dramatic and spectacular postures there is more room for exerting a more subtle and more durable influence on a selective basis in a number of areas. A piecemeal, evolutionary, path towards the attainment of desired objectives in the sphere of economic, political and social reform is the secret of the dynamism and resourcefulness of Western societies. It should also become their preferred tool in the sensitive area of East-West relations.

Notes

1 Sutton, op.cit., (chapter 2, note 5), vol.3, p.401.
2 Amann, in Amann, Cooper, Davies, op.cit. (chapter 2, note 12), p.276.
3 V. Sushkov in *Ekonomicheskoe sotrudnichestvo stran chlenov sev*, no.3, 1977, p.58.
4 P. Hanson, 'International Technology Transfer from the West to the USSR' in *Soviet Economy in a New Perspective*, JEC, 1976, p.797.
5 D.W. Green and H.S. Levine, 'Implications of Technology for the USSR' in *East-West Technological Co-operation*, NATO Colloquium 1976, Brussels, 1976, p.43.
6 Green and Levine, 'Macroeconomic Evidence of the Value of Machinery Imports to the Soviet Union' in Thomas and Kruse-Vaucienne (ed.), op.cit. (chapter 1, note 19), p.395.
7 Hanson, 'International Technology Transfer from the West to the USSR' (above, note 4), p.800.
8 J.P. Hardt, in *Technology Transfer and Scientific Co-operation between the US and USSR,* US Government Printing Office, Washington, 1977, p.81.
9 G.D. Holliday, *Western Technology Transfer to the Soviet Union: Its Impact on Soviet Resource Allocation*, Congressional Research Service, Washington, April 1978, p.3. Holliday's information is derived from I.U. Edwards, 'Automotive Trends in the USSR' in *Soviet Economic Prospects for the Seventies*, JEC, 1973, p.305; and Chase World Information Corporation, *Kamaz, the Billion Dollar Beginning*, 1974, p.66.
10 Ibid.

11 The efficiency with which an imported plant is run after handover is not the concern of the transferor of technology. His responsibility is to complete the 'guarantee trials' satisfactorily. Unfortunately little is known in detail about plant efficiencies in the USSR or in Eastern Europe in terms of Western measurements. Where plants are revisited some years later, they often provide the impression that high standards have not been maintained. According to E. Manevich, *Voprosy ekonomiki*, 1969, no.10, pp 27-40, a chemical plant imported from the West (put up at Shchekino) employed up to 8 times more workers than an identical plant situated in the West. For the category of white-collar employees the figure stood at about 3½.

12 P. Hanson, 'International Technology Transfer from the West to the USSR' (above, note 4), p.800.

13 O.K. Denisov, loc.cit. (chapter 3, note 17), based on table 2, p.39.

14 *EcoTass*, November 1977, no.45, p.5.

15 Rushing, loc.cit. (chapter 3, note 3), table 6, p.543.

16 P. Hanson, 'International Technology Transfer from the West to the USSR' (above, note 4), p.789.

17 E.A. Hewett, 'Recent Developments in East-West European Economic Relations and their Implications for US-East European Economic Relations' in *East European Economies Post-Helsinki*, JEC, 1977, p.181.

18 I. Dobozi, 'Problems of Raw Materials Supply in Eastern Europe', *World Economy*, January 1978, pp 205-22.

19 See, for example, CIA, *Soviet Economic Problems and Prospects*, Washington, 1977.

20 J. Berliner, *Prospects for Technological Progress*, JEC, 1976, p.434.

21 Duplication is also a very slow process and for this reason Khrushchev abandoned it in the early 1960s as the main strategy of catching up with the West. See Z. Medvedev, op.cit. (chapter 2, note 31), p.113.

22 For a recent account of the changes in the area of East-West commercial relations see J. Wilczynski, *The Multinationals and East-West Relations*, Macmillan, London, 1976.

23 L. Goure et. al., *The Convergence of Capitalism and Communism: the Soviet View*, Miami, 1973, p.v.

24 J. Wilczynski, *The Multinationals and East-West Relations* (above, note 22), pp 186-91.

25 See, for example, P. Wiles, 'On the Prevention of Technology Transfer' in NATO Colloquium 1976, Brussels, 1976, p.29.

26 Sutton, op.cit. (chapter 2, note 5), vol.3, chapter 27.

APPENDICES

Appendix A

A Comecon: basic information

The Council of Mutual Economic Assistance (CMEA) or Comecon was founded in January 1949 in Moscow. This move is usually interpreted in terms of Stalin's desire to respond to the American-inspired 'Marshall Plan' and also as a riposte to the founding of the Organisation for European Economic Co-operation. The founder members of Comecon are Bulgaria, Czechoslovakia, Hungary, Poland, Romania and the Soviet Union. The German Democratic Republic joined the group in 1950, Albania was a member in the period 1949-62, Mongolia became a full member in 1962 and Cuba in 1972 after a period of participation in Comecon bodies. Other socialist countries participate as observers. These include the People's Republic of Korea, Laos, China (temporarily between 1956 and 1966) and more recently Angola. Vietnam became a full member in 1978, and Finland and Yugoslavia have enjoyed a loose status of associate with Comecon since the mid-1960s.

Comecon is a result of exporting the Soviet politico-economic management into Eastern Europe, and at the time of its foundation it was simpler to transfer the Soviet structure in its totality to the new partners without making adjustments for disparate levels of development and national and cultural differences characteristic of the region. Soon after the departure of Stalin most of the Comecon countries began to react against the most harmful consequences of the rigid imposition of the command economic system by introducing reforms which involved a redistribution of decision-making power in favour of the enterprises (or enterprise associations) and introduction of market-based tools for directing the economy at the expense of purely administrative means. All Comecon members, however, stopped (or were compelled to stop) short of implementing a truly fundamental reform with the result that all the countries, with the possible exception of Hungary, may be said to be running their economies in accordance with the principles set by the Soviet model introduced in the 1950s, despite the indisputable fact that in some respects the command system has been modernised and made more efficient. If Comecon is an international extension of the Soviet system then any judgement made about the latter in the present study applies in greater or lesser degree to the other socialist countries. This is not to deny that in many respects, e.g., in the ability to generate and diffuse technological change, there exist significant inter-country differences. A detailed country by country examination, however, is beyond the scope of this study.

B Comecon: major compensation agreements

The following information relates to the discussion in chapter 5 of this study and should be regarded as a continuation of table 25.

Company	Year signed	Description of contract	Value	Products to be taken back	Duration of buy back	Marketing by	Annual value of chemical import
BULGARIA							
Technip/ Technipetrol, France/Italy	1972	Supply by Technipetrol/KTI of a 250,000 ton/year ethylene plant for Burgas	$120m. (investment)	25-30% of total in buy back. Only Technip will take chemicals including monoethylene glycol. Other chemicals are subject to availability. Also equipment and forklift trucks. Trucks started arriving 2 years ago. Technip and Balcan cars have formed a joint venture in France, Technocar. The Technipetrol j.v. is called Sibicar	Plant not ready yet. MEG is expected to come in from 1979 at a rate of 6,000 ton/ year over 4/5 years		$ 2.2m.
	1973	Supply by Technipetrol to Devnya of a 150,000 ton/year VCM plant based on PPG technology	$20m. (investment)				
	1974	Supply by Technip of a 50,000 ton/yea ldPE plant for Burgas based on the AT[illegible] process	FF100m. ($21.3m.)				
	1975	Supply by Technip of an 80,000 ton/year ethylene oxide plant based on SD technology for Burgas	FF190m. ($40.4m.)				
Klöckner/Huls West Germany	1974	Installation of 120,000 ton/year PVC plant at Burgas	DM44m.	100 per cent compensation in PVC. Hüls will market material in Europe. About 5,000 ton/ year	1980/81 for duration of credit 7-8 years		$ 3.5m.
Lurgi West Germany	1976	Supply of an 80,000 ton/year polypropylene plant at Burgas based on Hercules process		Have promised to take some Bulgarian goods. Chemicals will include pesticides, benzene, xylenes. Lurgi could use caustic soda in parent company but Bulgaria could not deliver		marketing through traders	difficult to assess
Haldor Topsøe, Denmark	1978	Supply of a 400 ton/day ammonia plant to be located at Vratsa	$10m.	Repayment in chemicals over 7 years. Ammonia, urea and other products to be included. Some for use in company plants, rest for sale in Scandinavia and elsewhere by Haldor Topsøe	1978 to 1984	Topsøe	over $ 1m.
Technipetrol, Italy	1978	Installation of a 200,000 ton/year urea plant at Vratsa	$16m.	Repayment in equipment and fork lift trucks			none

Company	Year signed	Description of contract	Value	Products to be taken back	Duration of buy back	Marketing by	Annual value of chemical import
UNDER DISCUSSION							
Furfural alcohol project		Installation of a 5,000 ton/year furfural alcohol plant at Vratsa	$4m.	Discussions are with Guiness & Peat of the UK. Repayment in chemicals including furfural alcohol		Lewis & Peat would market in the West	$ 1m.
CZECHOSLOVAKIA							
Chisso, Japan	early 1972	Installation at Zaluzi of an 80,000 ton/year polypropylene plant using Amoco process		C. Itoh offered to take 10% of capacity for sale (8,000 ton/year). Deliveries have begun		C. Itoh	$ 5m.
Mitsubishi Petrochemical, Japan	1977	Supply of an acrylic acid plant	$19m.	20% of contract value in counter purchase. No acrylic acid involved. Mitsubishi Corp. which will handle the deal has yet to come up with a proposal on which products it would like		Mitsubishi Corp.	$ 1m.
UNDER DISCUSSION							
Polypropylene project		Installation at Bratislava of a 30,000 ton/year polypropylene plant. Three teams are competing: Mitsubishi Petrochemical (Nissho Iwai would handle buy back), Chisso (C. Itoh), and Mitsui Toatsu (Sumitomo)	$30m.	Counter purchase is not fixed yet			about $ 1m.
EAST GERMANY (GDR)							
Uhde/Hoechst, West Germany	1976	Installation on a turnkey basis by Uhde at the Buna works in Schkopau of a complex comprising plants for 200,000 ton/year chlorine, 220,000 ton/year caustic soda, 200,000 ton/year VCM, and 100,000 ton/year PVC	DM1 billion ($472m.)	Refinance to the value of 25-30% in PVC and caustic soda. The PVC part is less than 15,000 ton/year and the caustic soda supplies will be over 30,000 ton/year. Some caustic soda will be	1980 to 1987	Hoechst	$10m. $ 4m. $14m.

Company	Year signed	Description of contract	Value	Products to be taken back	Duration of buy back	Marketing by	Annual value of chemical import
				used in Hoechst plants, but all the PVC and the major part of soda will be incorporated into the Hoechst marketing strategy, including the West European markets.			
KTI/ The Netherlands	March 1977	Supply of one furnace for the Böhlen cracker	DM12m. ($5.7m.)	50% compensation. Bidding started with 20%, went up during negotiations. Of the total: 35% in chemicals the rest in equipment. Chemicals include potash, but difficult to market because GDR has own territory where contractor is not allowed to sell. Middle East mainly. Supplies have not been reliable. Contract will involve roughly 10,000 ton/year of KCl	March 1977 to 1980	markets through Cheminter, part of Guiness & Peat	terminates in 1980
Catalytic, UK	1977	Installation of 130,000 ton/year chlorine plant at Bitterfeld based on Diamond Shamrock technology	£20m. ($36m.)	Compensation in 20,000 ton/ year caustic soda over 5 year period			$ 2.3m.
Mitsui, Japan	Nov. 1977	Installation of an aromatics complex at Schwedt centred on a 250,000 ton/year benzene plant	$380m.	Original agreement is listed here but because of Dow's tolling arrangement (see below) this has changed. a) a total of 265,000 ton of benzene over 6 years (1983-88) b) a total of 338,000 ton of propylene over 7 years (1982-88) c) C_8 fraction over 6 years at a rate of 10-20,000 ton per year (1983-88) Propylene is now out. Rest marketed by Brenntag	1982 to 1988	Brenntag	$25m.

Company	Year signed	Description of contract	Value	Products to be taken back	Duration of buy back	Marketing by	Annual value of chemical import
Dow Chemical Europe (tolling arrangement)	1977	Signed an exchange agreement with AHB Chemie to supply the polyols complex at Schwarzheide with propylene oxide from the Stade plant near the border. Stade plant has a 240,000 ton/year capacity	$250m.	In exchange Dow will receive propylene from the Schwedt plant (about 50,000 ton/year). Partly for use at Stade, rest for sale. The Terneuzen plant supplies the bulk of the propylene required at Stade. The agreement works out roughly $165m. to E. Germany and $85m. from there	1979 to 1988		
UNDER DISCUSSION							
Fertiliser project		Installation of a calcium ammonium nitrate complex at Rostock to produce 1.1m. ton/year	$200m.	Repayment in 140,000 ton per year of calcium ammonium nitrate over 10 years, beginning in 1984	1984 to 1993		$14m.
Polyester fibre project		Installation of plant at Pima to produce 10-12,000 ton/year technical polyester yarn	DM200m. ($100m.)	50% buy back at between 4-6,000 ton/year. Discussions are with Zimmer of West Germany which would market through specialists in North and South America	1982 to 1988		$ 9.25m.
HUNGARY		Hungary has some compensation agreements, notably in PVC, but no information is available					
POLAND							
Krupp Koppers/ Kronos Titan, West Germany	1972	Supply of a 40,000 ton/year titanium dioxide plant located at Police. The plant is already on stream, working at 50% capacity. It uses Kronos technology		For the plant both Krupp and Kronos received a fee, but the agreement contained an option for Kronos to purchase 2-3,000 ton/year to titanium dioxide. So far the 8 year option has not been exercised		If it is taken up Kronos will market	$ 2-3m.

Company	Year signed	Description of contract	Value	Products to be taken back	Duration of buy back	Marketing by	Annual value of chemical import
Voest Alpine/ Chemie Linz, Austria	1974	Construction by Voest Alpine of a 32,000 ton/year melamine plant at Pulawy based on Chemie Linz know-how. Commissioning in third quarter of 1978. Will be running at 70-75% of capacity	A.Sch 800m. ($53m.)	Chemie Linz will take back products to the value of its licence which is around 10-15% of total. According to Chemie Linz this equals a total of 10-15,000 ton of melamine. In fact Poland will probably sell most of additional production to USSR and Romania, so that Chemie Linz will take other products such as phosphoric acid and sulphur. The melamine that does come back will not be marketed in Europe, USA, Canada or Brazil	1979 to 1988	traders to market in Far East, Africa	$ 1m.
Petrocarbon Developments, UK	1975	Installation at Wloclawek of a 200,000 ton/year PVC plant based on Shinetsu technology. Also 180,000 ton/year chlorine plant to be installed by Catalytic based on Diamond Shamrock process and a 205,000 ton/year VCM plant to be built by Davy Powergas using PPG Industries technology	£125m. ($225m.)	Negotiations began with compensation, eventually this was dropped but Poland earmarked 40,000 ton/year of PVC to sell to the West. The country also has spare caustic soda. Some of the chlorine from Rokita plant at Brzeg Dolny (140,000 ton/ year capacity), built by Uhde, has also been earmarked for export by Ciech	1980-1989	Ciech to market	$27m.
Krebs/Klöckner France/West Germany	1975	Installation of a 460,000 ton/year soda ash plant at Inowroclaw	FF850m. ($182m.)	During negotiations the contract was based on compensation. Later Poland dropped this idea and decided to market itself. Klöckner will help market some material. Total sales to the West work out at some 250,000 ton per year of soda ash		Ciech to market most	$30m.

Company	Year signed	Description of contract	Value	Products to be taken back	Duration of buy back	Marketing by	Annual value of chemical import
Creusot Loire, France	May 1976	Installation at Police of a fertiliser complex comprising two 750 ton/day ammonia plants and one 1,200 ton/day urea unit, also 2,600 ton/day NPK plant. Heurtey designed both ammonia plants based on Haldor Topsøe technology. The urea unit uses Stamicarbon technology. Foster Wheeler France responsible for NPK plant based on Fisons know-how	FF1,800m. ($391m.)	100% compensation to be repaid in urea at a rate of 400,000 ton/year over 10 years. Under the contract deliveries should start in 1981 but plant is late and Poland asked from 1983. Poland trying to offer other products because it needs the urea	1983 to 1992	to be marketed by Interore of Rome, Occidental's trading arm	$52m.
Occidental Petroleum, USA	1978	Signed a 20-year swap deal to supply Poland with 1m. ton/year of phosphate rock in exchange for 500,000 ton/year of sulphur. Oxy will help finance a new 500,000 ton/year sulphur mine in Poland	$1.33 billion	Occidental will use the sulphur in its US operations including the production of superphosphoric acid for its supply deal with the USSR	1978 to 1997		$33.5m.
UNDER DISCUSSION							
Coal gasification project		Coal gasification project which will supply synthesis gas for 4 plants: 500,000 ton/year ammonia unit, two 500,000 ton/year urea plants and a 130,000 ton/year oxo alcohols unit. Of all the projects under discussion this one is the most likely to go ahead because Poland has already been offered credit from Germany on which it has to make a decision by October	$500m. for gasification and chemicals plant				$50m.
Chloroprene rubber project		Installation of a 50,000 ton/year chloroprene rubber plant at Oswiecim, will be designed for easy expansion to 100,000 ton/year. Also a 160,000 ton/year chlorine plant will be built	$200m.	No decision has yet been made and it is likely that this project will be postponed. Eventually however it will also have to be compensated			$25m.

Company	Year signed	Description of contract	Value	Products to be taken back	Duration of buy back	Marketing by	Annual value of chemical import
Polyester fibre project		Setting up of a grass roots polyester fibres complex at Pila with capacity of 67,000 ton/year also a 120,000 ton/year DMT plant and a 72,000 ton/year benzene unit at Plock. (Due to its financial difficulties Poland is thinking of postponing this investment)	about $500m.				$64m.

Products offered by Poland in return in 000t

	1982	1983	1984	1985	1986	1987	1988	1989	1990	1991	Total
DMT	50.3	36.5	16.5	16.5	16.5	16.5	16.5	16.5	15.5	16.5	218.8
paraxylene	30.5	15.0	15.0	15.0	15.0	15.0	15.0	15.0	15.0	15.0	165.5
benzene	68.1	72.0	72.0	72.0	72.0	72.0	72.0	72.0	72.0	72.0	716.1
methanol	100	100	100	100	100	100	100	100	100	100	1m.
polyester film	1.0	1.0	1.0	1.0	1.0	1.0	1.0	1.0	1.0	1.0	10.0
polyester chips	–	7.0	10.0	10.0	10.0	10.0	10.0	10.0	10.0	10.0	87.0
polyester filament yarn	3.4	9.7	10.8	9.7	8.5	7.3	6.2	5.0	3 9	2.8	67.3
polyester fibre	7.5	–	–	–	–	–	–	–	–	–	7.5

Company	Year signed	Description of contract	Value	Products to be taken back	Duration of buy back	Marketing by	Annual value of chemical import
ROMANIA							
General Tyre, USA		Supply of know how for a rubber tyre plant under the Victoria name at Floresti. $19m. in compensation		Moody International of Austria was to have taken the compensation in chemicals. To date nothing has been taken	to 1982	Arcode may be dealing with this	flexible based on $19m. total
Linde, West Germany	1973	Undisclosed contract		DM10m. worth of compensation in chemicals, to be completed immediately			
KTI, The Netherlands	1978	Supply of a CO Cosorb unit for the acetic acid plant at Craiova	*f*5.5m.	25% in counter purchase in textiles, will also take some urea			small

Company	Year signed	Description of contract	Value	Products to be taken back	Duration of buy back	Marketing by	Annual value of chemical import
DeNora, Italy	1978	Installation of two 85,000 ton/year chlorine plants	$18m.	Counter purchase involved			
UNDER DISCUSSION							
Titanium dioxide project		Installation of a plant in Romania to produce 20,000 ton/year titanium dioxide. Negotiations are with Sim-Chem offering American Cyanamid technology		About 30% counter purchase will be involved in chemicals		TR International interested in marketing	
YUGOSLAVIA							
Klöckner/Davy Powergas GmbH West Germany	1976	Installation of a 20,000 ton/year dioctyl phthalate plant for Inex Bixol at Vrsac	DM10m. ($5m.)	Repayment with 8,000 ton/year of DOP over six years beginning in 1978/79	until 1984	Klöckner to market in world markets including W. Europe	$ 4.3m.
Voest Alpine, Austria	1977	Supply for Azot Gorazde of a 160 ton/day concentrated nitric acid and a 160 ton/day ammonium nitrate plants on Davy Powergas and Kaltenbach technology respectively	A.Sch 200m. ($13.3m.)	Parallel contract signed with Feroelectro, a trading company of Sarajevo for counter purchase to the value of ASch60m. Until 1979. A third of this is in chemicals	1977 to 1979	Voest Alpine Industrial Trade Co.	terminates in 1979
Foster Wheeler, UK	1977	Supply of an aromatics complex for Ina at Sisak based on UOP technology. It will comprise units for 100,000 ton/year benzene, 80,000 ton/year toluene, 73,000 ton/year paraxylene and 55,000 ton/year orthoxylene	$17.5m.	30% of contract value in counter purchase, i.e. $5.25m. Negligible amount of chemicals mainly aluminium sulphate and copper sulphate		Foster Wheeler	small
Foster Wheeler, UK	1977	Installation of a catalytic cracker for Naftagas	£3.3m. ($6m.)	30% in counter purchase about $2m.			negligible

Company	Year signed	Description of contract	Value	Products to be taken back	Duration of buy back	Marketing by	Annual value of chemical import
Litwin, France	1977	Supply of carbon black plant for Ina at Kutira. Total capacity will be 40,000 ton/year (expansion of old 18,000 ton/year to 40,000 ton/year and installation of new 20,000 ton/year plant)	$10m.	Will purchase Yugoslav products to the value of $5m. within a five year period, maybe some chemicals too	1978 to 1982	A subsidiary Amtel Chimie Plastique will market	small
Creusot Loire/ Kellogg, France/UK	Dec. 1977	Installation of 1,360 ton/day ammonia and 1,500 ton/day urea plants. Ammonia plant to be engineered by Creusot Loire based on Kellogg technology. Kellogg Continental of Amsterdam will engineer urea plant using Stamicarbon process	$85m.	50% of contract value to be compensated but the list of products is open. Some fertilisers will be included. Over 5 years	1978 to 1982		$ 5m.
Davy Powergas, UK	1977	Installation of a fertiliser complex for Ina at Kutina to produce 500,000 ton/year sulphuric acid, 165,000 ton/year phosphoric acid, 150,000 ton/year nitric acid, 165,000 ton/year ammonium nitrate and 500,000 ton/year MAP and NPK fertilisers	£36m. ($65m.)	Counter trade to the value of 30% in Yugoslav goods. Probably no chemicals	over 3 years		
Klöckner Ina/ KHD, West Germany	end 1977	Supply of a 10,000 ton/year maleic anhydride plant for Koksno Hemijski Kombinat Boris Kidric to be located at Lukavac, using Scientific Design technology. For completion spring 1979	DM18m. ($8.5m.)	30% of contract value to be repaid in maleic anhydride (total of 7,000 ton)	1980 start for 3 to 7 years	Klöckner	$ 2m.
Sim-Chem, UK	1977	Supply of a 50,000 ton/year suspension PVC plant for Jugovinil at Split, based on BF Goodrich technology	$25-30m.	30% of contract value to be repaid in PVC (possibly other products)		BSE Genex (Yugoslavia's major trading co.)	$ 1m.
Ingeco Laing, UK	1978	Supply by Ingeco Laing of a 50,000 ton/year linear alkyl benzene plant for Prva Iskra at Baric	$20m.	$6m. of buy back to include chemicals		third party (undisclosed) to market	$ 1m.

Company	Year signed	Description of contract	Value	Products to be taken back	Duration of buy back	Marketing by	Annual value of chemical import
Badger, UK	1978	Installation for Ina at Sisak of a 40,000 ton/year SBR plant	$17.5m.	30% of contract value in Yugoslav goods, i.e. $5.25m.			$ 1m.
UNDER DISCUSSION							
Fertiliser project		Installation of a nitrophosphates complex for RTB Bor at Subotica to produce 700 ton/day nitric acid and 800-1,500 ton/day NPK fertilisers and 500-700 ton/day ammonium nitrate	$85m. investment cost	Successful bidder will have to take 30% of contract value in Yugoslav goods			about $ 4m.
Synthetic rubber project		Installation for Fabrika Sintetikog Kaucuka (FSK) of a 40,000 ton/year SBR plant and a butadiene extraction unit to be located at Zrenjanin					

Source: *European Chemical News*, European Review, 21 July 1978.

Appendix B

Major projects in petrochemicals and related branches recently completed or currently under construction

The map showing the Soviet petrochemical centres, to which the following list relates, is divided into ten economic regions. The information is, therefore, divided into ten sections (A, B, C, D, E, F, G, H, J, K), each illustrating the types of plants which were, or are being, constructed at specific locations falling within a particular region.

It should be noted that more detailed information on some of the more recent cases of technology transfer involving 'buy-back' agreements is produced in chapter 5, table 25. The following data are concerned mainly with the geographical distribution of Western plants in the USSR.

Region A includes the following republics: Lithuania, Latvia, Estonia, Belorussia, Ukraine and Moldavia.

Location	Company	Product	Capacity	Date of completion
Chernigov	High Polymer and Petrochem Eng.	Nylon 66 tyre cord		1965
	Zimmer & Vickers	Nylon 66		1966
	Zimmer & Vickers	Nylon 66		1974
Dneprodzherzhinsk	Creusot-Loire	Phosgene and polymethylene polyphenyldi-isocyanate		1974
	Creusot-Loire	Polyurethane		1977
	Litwin	Isocyanates	22,000 tpa	1978-79
Donetsk (Rubezhnoe)	Davy Powergas and Klöckner	Pthalic anhydride	60,000 tpa	1980
	Ind. Anlagen	Malleic anhydride	3,000 tpa	
Gorlovka	Davy Powergas	Polystyrene		1962
Kalush	UHDE/Goodrich – Hoechst process	PVC	250,000 tpa	1975

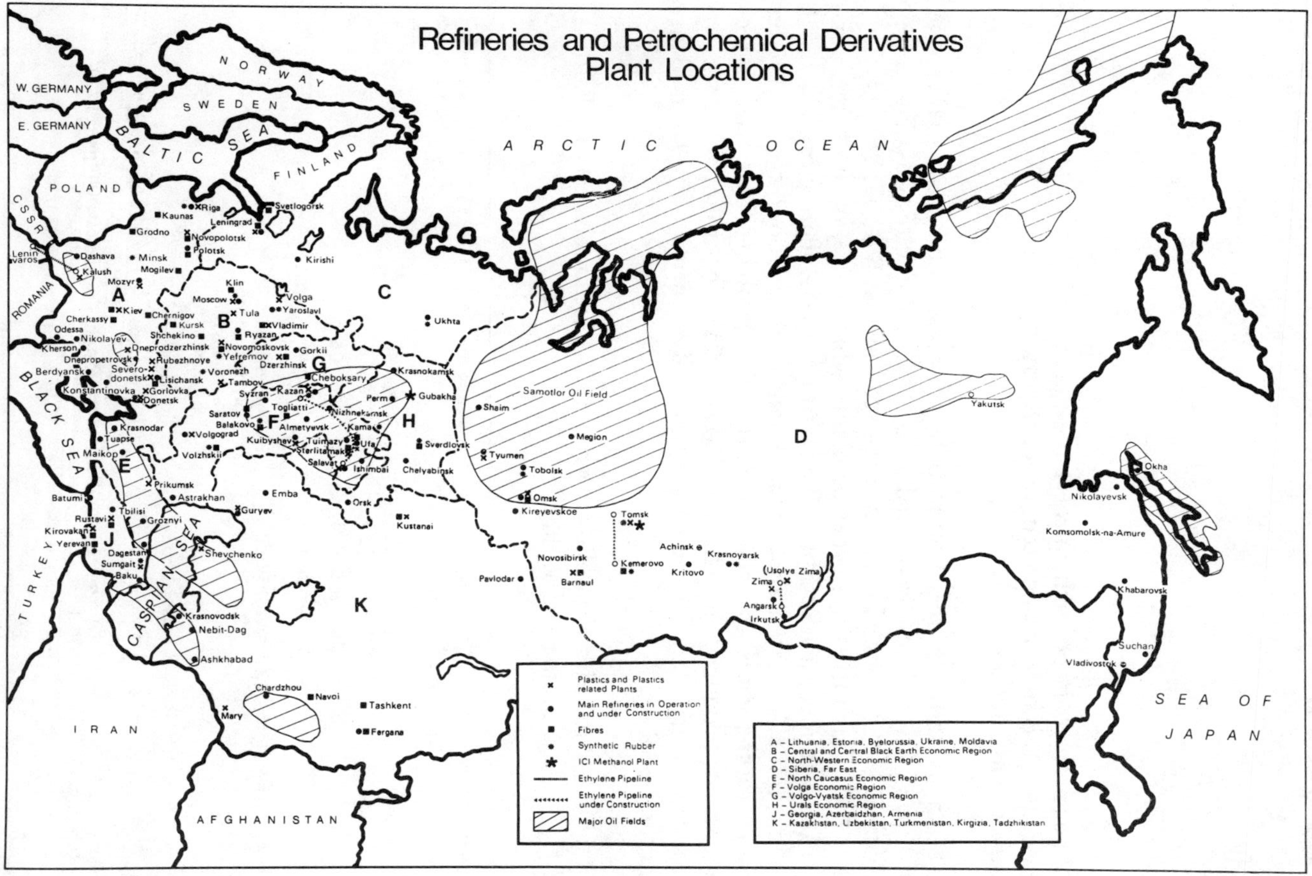

Refineries and Petrochemical Derivatives Plant Locations

Between 1972 and 1975 a 300 km long ethylene pipeline was constructed by Petrole-Chimie Eng. with an annual throughput of 140,000 tons. The pipeline links Kalush with a new naphtha cracker at Leninvaros in Hungary constructed by Linde. The complex produces 250,000 tpa ethylene and 130,000 tpa propylene. Part of the Hungarian ethylene is transported to Kalush via the pipeline, plus 80,000 tpa propylene by rail. The USSR pays Hungary by deliveries of HDPE, LDPE, polystyrene, acrylonitrile, ethylene glycol and styrene.

Location	Company	Product	Capacity	Date of completion
Kaunas	Courtaulds	Cellulose triacetate fibre		1965
Kiev	Maurer	Carbon disulphide		1962
Lvov	SNIA	Carbon disulphide		1974
Minsk	Krupp	Tyres		1977
Mogilev	Polyspinners (ICI, CJB, P. Brotherhood)	Polyester staple fibre	20,000 tpa	
		Polyester staple and filament	30,000 tpa	1970
	Krupp	DMT		1975
	Krupp	DMT	60,000 tpa	early 1980
	Krupp	DMT	120,000 tpa	early 1980
	Hoechst/UHDE	Polyester staple fibre expansion	35,000 tpa	1980-81
	Dynamit-Nobel process	Polyester fibre	55,000 tpa	1980-81
	Sim. Chem	DMT	6,000 tpa	1976
	CJB, ICI, P. Brotherhood	Polyester staple expansion	10,000 tpa	1979
Mozyr	Asahi	Benzene	120,000 tpa	early 1980s
Novopolotsk	Humphreys and Glasgow	Ethylene		1968
	Marubeni	Acrylonitrile	50,000 tpa	1968
	Marubeni	Acrylonitrile	25,000 tpa (extension)	1978
	Simon Carves	LDPE		1968
	CJB	P-Xylene		1971
	Courtaulds	Polyacrylonitrile fibre		1972
	Asahi	Benzene	120,000 tpa	early 1980s
Severodonetsk (Lisichansk)	BASF	Acetylene		1963
	High Polymer and Petrochem Eng.	A.H. Salt (for nylon 66)		

Location	Company	Product	Capacity	Date of completion
Severodenetsk (cont.)	Zimmer	PV pyrrolidene & methypyrrolidine		1964
	Linde	Ethylene	250,000 tpa	
		Propylene	125,000 tpa	
		Benzene (the complex will eventually also turn out: high grade petrol, styrene monomer, ethylbenzene, sulphur diesel oil)	100,000 tpa	1977
	Salzgitter	LDPE	240,000 tpa	1978

Region B consists of the Central and Central-Black-Earth economic regions. The most important recent Western plants include:

Location	Company	Product	Capacity	Date of completion
Kursk	Krupp	Polyester fibre	17,500 tpa	1961
	von Kohrn	Polyester fibre	2,450 tpa	1961
	Krupp	Polyester fibre	3,000 tpa	1967
	Hoechst	Polyester fibre	55,000 tpa	1980
Moscow	Anger	PVC	50,000 tpa	1966
	Serez	Laminated plastic panels		1967
	Plane Trade	2 x polymethylmethacrylate		1968
	Japan Steel	Polyester film		1970
	Alkor & Krauss Maffa	PVC		1970
	Petrole Chemie	Paracresol		1973
	Pirelli	Tyres		1977
	Lurgi Apparate Technik	6 degumming plants	100 tp/day each	1980-81
	Linde	Ethylene	250,000 tpa	early 1980s
		Propylene	125,000 tpa	early 1980s
		C4 fraction	70,000 tpa	early 1980s
		Benzene	77,000 tpa	early 1980s
		Benzole	100,000 tpa	early 1980s
Novomoskovsk	Krupp	DMT		1960
	Snamprogetti	Methanol		1964
	Krupp	DMT		1966
	ERI	Acetylene		1966
Ryazan	von Kohrn	Viscose rayon		1960
Shchekino	Montecatini	Methanol		1966

Location	Company	Product	Capacity	Date of completion
Shebekino	CJB	Fatty acids		1964
Tambov	Montecatini	Malleic anhydride		1963
	UCB	Acrylic resin		1970
	Montedison	Dyestuff complex	4,000 tpa	1979
Tula	Marubeni Iida	Plastics (unspecified)		1971
Vladimir	Sim-Chem/ICI	Polyester film	5,000 tpa	1976
Yaroslavl	Pirelli	Tyres		1964
	CJB	Solid latex		1972
Zagorsk	Montecatini	Dyestuffs		1968
Zhilevo	Kestner	Polyester resins		1962

Region C is the North-Western region of the European USSR. Important locations in this region are:

Leningrad	Plane Trade	Polymethyl-methacrylatenoly and styrene	1968
	Kawasaki and Atlantic Richfield	Xylenes (O and P)	1974
Svetogorsk	Rauma Repola	Cellulose acetate	1977

Region D consists of Siberia and the Soviet Far East.

Development of Siberia and the Far East is correctly perceived by Soviet planners and leaders as the key precondition for further growth of the economy. The region is notorious for vast deposits of raw materials and minerals of all kinds. However, the extraction and transport of these raw materials pose considerable problems owing to unfavourable climatic conditions and the large distances involved.

Siberia is rich in oil, natural gas and coal, and the successful implementation of Soviet energy policy is becoming increasingly dependent on utilisation of these resources. At the same time, it is more and more being recognised that Siberia is especially suitable for the development of the chemical and petrochemical industries owing to its raw material, cheap energy and water endowment. Another advantageous factor is that the chemical industry is relatively less labour-intensive; it seems that the most difficult problem faced by the planners is to ensure the supply of the labour force in this inhospitable region.

The awareness of the urgent need to develop the Siberian oil fields dates back to the early 1960s, although the first oil refineries began to be constructed there in 1955. But the construction of large petrochemical complexes in Siberia started relatively recently and there have been a number of controversial issues to be resolved. In 1974, however, a decision was reached at the Central Committee level to go ahead with the development of several chemical centres and, as usual, co-operation of Western companies was encouraged.

The following information illustrates the scope of those efforts and of the Western involvement in it.

Location	Company	Product	Capacity	Date of completion
Barnaul	Zimmer	Nylon 6		1960
	Maurer	Cellophane		1961
	Lurgi	Acetate fibre	50,000 tpa	1975
Irkutsk	A West German firm	P-xylene		1966
Omsk	Litwin	High impact polystyrene	100,000 tpa	1978
	Technip	Raw naphtha based complex:		1980
		Benzene	125,000 tpa	
		O-xylene	165,000 tpa	
		P-xylene	165,000 tpa	
	Euro-Technica delivers the xylene separation units			

Tobolsk is intended to become a major synthetic rubber centre. The Russians plan to manufacture the following products there: ethylene, polyethylene, polypropylene, butadiene, isoprene, benzene, methanol and other products.

The construction at Tobolsk started in 1972. Relatively greater use is made of Soviet technology and know-how and some of the processes used will be Soviet designed. The Russians, however, have been reported to be aware that their processes are relatively complicated and more expensive and they are attempting to improve them.

In spite of this, Western companies will play a significant part in the development of this centre, although at the time of writing no firm indications as to what will be constructed there by Western companies are available. The most recent news indicated that the Russians will soon invite offers for bids for construction of a 200-300,000 tpa ethylene oxide plant to be built at Tobolsk. Scientific Design/Salzgitter are considering this opportunity whilst Litwin may get the polystyrene plant.

Tomsk: German Consortium. Krupp-Koppers, Linde, Klöckner, Uhde, Salzgitter are currently studying Sovict proposals for a massive S4 billion complex based on a 10m. tpa oil refinery. There are different versions of the project but it will probably take the following shape:

ethylene	900,000 tpa
propylene	450,000 tpa
benzene (toluene, xylenes)	balance
styrene	600,000 tpa
polypropylene	400,000 tpa
LDPE, HDPE	400,000 tpa
vinylchloride	320,000 tpa
polyvinylchloride	250,000 tpa
dichloroethane	320,000 tpa
ethylene oxide	300,000 tpa

(This information refers to the original proposal which has been scaled down. See table 25 and discussion in chapter 5.)

Krupp are acting as consortium leader. Linde will get the olefins units, aromatics and the dichloroethane plant. Salzgitter will get the LDPE and HDPE plants and the ethylene oxide plant. Uhde hopes to get the VCM plant order using Goodrich process and Klöckner will probably secure the PVC plant with KHD as subcontractor offering Huls technology.

The French company, Litwin, though not a member of the consortium, will probably get the polystyrene order, and Montedison the polypropylene plant.

Also in Tomsk, ICI in conjunction with Davy Powergas and Klöckner are constructing a methanol plant: capacity 2,500 t/day; date of completion, 1980. Montedison are already constructing a polypropylene plant at Tomsk: capacity 100,000 tpa; date of completion 1980.

Angarsk: there is an ethylene plant: capacity 450,000 tpa; and a number of downstream units. The USSR is constructing two LDPE plants: combined capacity 120,000 tpa. This process, called Polimir 60, follows successful development of Polimir 50 at Novopolotsk, which was developed jointly by the USSR and GDR.

Zima (Usolye Zima) is a new site close to substantial salt deposits. The Russians want to produce acetylene, chlorine derivatives and PVC there, from ethylene supplied from Angarsk by a pipeline.

Uhde using Hoechst/Goodrich process are constructing a VCM plant: capacity 270,000 tpa; date of completion, 1979.

Klöckner, using Huls technology, are constructing a PVC plant: capacity 250,000 tpa; date of completion, 1979. Salzgitter are providing storage and packing facilities. Chlorine plant will be supplied by Hooker.

Angarsk-Zima pipeline: CJB are constructing a 230 km long 8 in diameter ethylene pipeline system between Angarsk and Zima. Soviet pipes are used and CJB are building two compressor stations. Annual throughput is 200,000 tpa. Eventually the Russians want another ethylene plant at Zima.

Region E consists of the North Caucasus economic region. Recent Western involvement is taking place in two locations:

Location	Company	Product	Capacity	Date of completion
Budyonnovsk (near Prikumsk)	CJB/Union Carbide technology used	HDPE	200,000 tpa	1979
Groznyi	Snamprogetti	LNG fractionation plant	200,000 ton per 3 days	1976

Region F is the Volga economic region, the traditional centre of the Soviet chemical industry. Western involvement in this region has been considerable. The recent examples are the following:

Balakovo	Chatillon	Viscose rayon tyre cord	1963
	Courtaulds	Viscose rayon tyre cord	1964
	Chatillon	Viscose rayon tyre cord	1966
	Pirelli	Rubber products	1970
Kazan	Salzgitter	LDPE	1966
	Simon Carves	LDPE	1968
	Humphreys and Glasgow	Ethylene	1968

Location	Company	Product	Capacity	Date of completion
Kazan (cont.)	CJB	Organic peroxide		1968
	Mitsui	Ethylene oxide Ethylene glycol		1969
	Sturtevant Engineering	Insecticides		1969
	Toyo	Ethylene	100,000 tpa	1973
	Salzgitter/Union Carbide	LDPE	120,000 tpa	1976
	Nippon Shokubai Kagaku Kogyo Comp	Ethylene oxide	60,000 tpa	1976
	CJB	HDPE	200,000 tpa	1981
Kazan-Nizhnekamsk pipeline	In 1976 CJB was awarded a £7.75m. order for the pipeline and compressor station linking Kazan with Nizhnekamsk, Ufa, Sterlitamak and Salavat. This pipeline will carry ethylene initially from Nizhnekamsk to Kazan.			
Nizhnekamsk	Lummus	Ethylene	450,000 tpa	1976
	Salzgitter	Ethylene oxide	200,000 tpa	1981
	Pressindustria	Ethoxylates surfactants recovery	250,000 tpa	1981
Kuybyshev	Salzgitter	HDPE		1962
	Humphreys and Glasgow	Phosphorus pentasulphide		1966
Naberezhnye Chelny (The Kama River Works)	Pirelli	Tyres		1977
Salavat	Salzgitter	LDPE		1962
Saratov	Courtaulds	Cellulose acetate	5-7,000 tpa	1962
	Courtaulds	Acrylic Staple	4,500 tpa	1963
	Montedison	Acrylonitrile	150,000 tpa	1970
Sterlitamak	Montecatini	Acetylene ethylene		1964
	Montecatini	Ethylenediamine		1965
	Wycon Services	DMEU		1965
	Montedison	Sodium chloride crystallisation	270,000 tpa	1977
Togliatti	IMI/Fiat	Fibres		1970
Ufa	Wycon Services	Herbicides		1964
	Degussa	Carbon black		1965
	Salzgitter	LDPE		1966
	Simon Carves	LDPE		1967
	Technip/Eurotechnica	O-xylene P-xylene Benzene	165,000 tpa 165,000 tpa 125,000 tpa	1980

Location	Company	Product	Capacity	Date of completion
Volgograd	Fuji and Fureha	PVC		1968
	Technimont	Chlorofluor-methanes	30,000 tpa	1970
Volzhskiy	Snia	Carbon disulphide		1970
	C. Itoh	Elastomeric fibre		1974

Region G represents the Volgo-Vyatsk economic region. Important olefin-derived and olefins industry centres in the region are:

Cheboksary	Montecatini	Dyestuffs		1968
Dzherzhinsk	Speichim	PVC		1968
Gorkii	Salzgitter	Ethylene oxide Ethylene glycol	200,000 tpa 120,000 tpa	1979

Region H is the Urals economic region. Western equipment deliveries falling within the scope of this section are the following:

Gubakha	Davy Powergas/ICI/ Klöckner	Methanol	2,500 tp/day	1981
Perm	Davy Powergas/GBHH/ Klöckner	Pthalic anhydride Fumaric acid	60,000 tpa 3,000 tpa	1980-81

The USSR is building a new petrochemical complex here. The first phase includes ethylene and propylene units.

Region J consists of the trans-Caucasian republics, Georgia, Azerbadjan and Armenia. At present about 90 per cent of the region's chemical production comes from the following complexes: Baku- Sumgait, Yerevan-Kirovakan and Rustavi.

The region has large deposits of oil, natural gas and other raw materials essential for production of chemicals such as salt, iodine and bromine. Recent Western involvement is recorded as follows:

Kirovakan	Courtaulds	Cellulose acetate fibre	6,800 tpa	1962
	Technimont	Melamine	10,000 tpa	1973

Location	Company	Product	Capacity	Date of completion
Rustavi	Snia	Caprolactam recovery plants		1964
Sumgait	Simon Carves	LDPE		1969
	Dshikawajima/Harima	PVC		1971
	Construction is currently under way in Sumgait of a 300,000-330,000 tpa ethylene plant, date of completion 1980. The plant is Soviet built, using Soviet technology. Downstream production will mainly focus on Epoxy and polyester resins.			
Yerevan	Spei	Cellulose acetate		1963
	Spei	Acetic anhydride		1963
	Power Gas/BP technology	Acetic acid		1968
		Chloroprene monomer (rubber)	75,000 tpa	1977
	Kobe Steel of Japan	Chloroprene monomer (rubber)	50,000 tpa	1980s

Region K consists of the following republics: Kazakhstan, Uzbekistan, Turkministan, Kirghizia and Tadjikistan. In recent years expansion of production facilities has been taking place at a number of locations, e.g. Chardzhov, Mary, Navoi, Kustanai, Schevchenko and Guryev. Our investigation shows the following Western involvement in this region:

Chirchik	Snia Viscosa	Caprolactam	80,000 tpa	1979
Fergana	Montedison	Cellulose triacetate	42,000 tpa	1977
Guryev	Montedison	Polypropylene, copolymers	30,000 tpa	1977
	–	HDPE	23,000 tpa	1966
	To be decided	HDPE expansion	to 35,000 tpa	1980s
Pavlodar	Asahi	Benzene	120,000 tpa	1980
Shevchenko	Litwin	Styrene	300,000 tpa	
		High impact polystyrene	100,000 tpa	
		Expandable polystyrene boards	4,400 tpa	1980
Tashkent	Snia Fiscosa	Caprolactam	80,000 tpa	1980
Yavan	Technimont	Chlorofluormethanes	30,000 tpa	1979

Appendix C

A Soviet ammonia capacity build-up

The relevant map shows all ammonia production locations in the USSR. The symbols indicate recently constructed or planned Toyo, Chemico or Creusot-Loire plants. Virtually all the plants built during this decade or the early 1980s will be based on natural gas and their capacity is 450,000 tpa (370,000 tpa N). An exception to this is the Toyo plant at Angarsk which will utilise naphtha. The only major ammonia project under completion reportedly based entirely on Soviet efforts is the site at Chirchik in Uzbekistan. The latter stage Toyo plants, to be on stream in 1981, will allegedly be using a Soviet process. Experts, however, believe that it is highly unlikely that a genuine Russian process is involved. Most likely these plants will use the Kellogg process, like most of the previous ones, with the difference that the Russians will have acquired the right to call it their own by the time the plants are in operation. An important feature of this later stage of the Soviet ammonia programme is that more and more Soviet or East European parts will be supplied. For example, Czechoslovakia will deliver the reformer sections for the plants. This equipment is thought to be a replica of Western equipment now manufactured in the USSR. The Russians themselves now possess a Western plant for manufacture of spare parts for the compressor section of ammonia technology. They hope eventually to use this capacity for construction of complete compressors which will be used in the later series of the planned plants.

The Russians have been reported to be developing their own technology for producing up to 3,000,000 tons of ammonia per annum. It is not, however, clear whether this would be a single unit or a composition of several lesser capacity units. In any case, Western experts suspect that such large plants would be uneconomical because of reduced flexibility. Nevertheless, they are technically feasible and apart from that the Soviet economic system generates an environment in which such large units may be economically justified.

A word of caution is due in regard to the fulfilment of the construction targets of the ammonia plants. Some experts believe that the plants cannot be on stream by the dates stipulated in the plan (see below). The whole programme could be slowed down by several years and thus

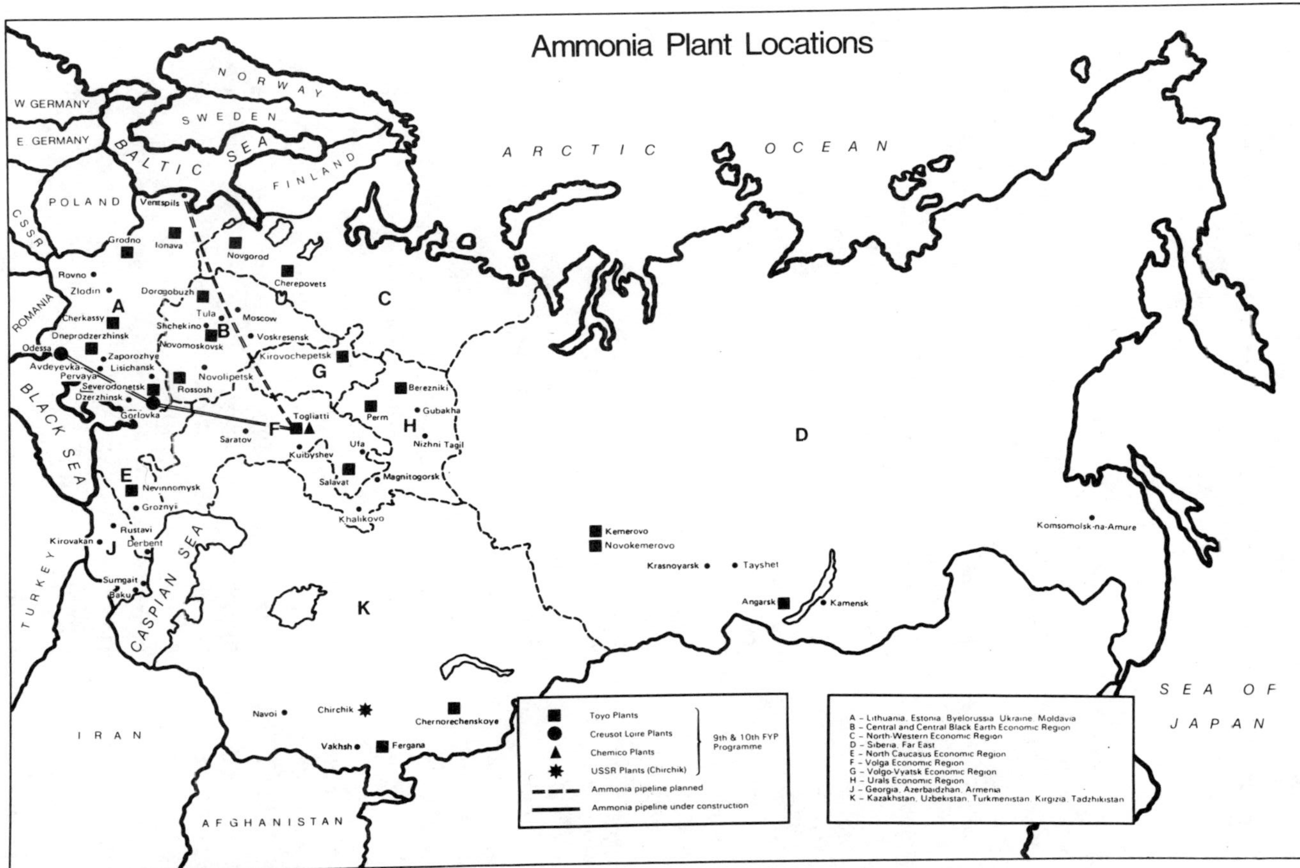
Ammonia Plant Locations
W GERMANY
E GERMANY
NORWAY
SWEDEN
FINLAND
BALTIC SEA
POLAND
CSSR
ROMANIA
BLACK SEA
TURKEY
IRAN
AFGHANISTAN
CASPIAN SEA
ARCTIC OCEAN
SEA OF JAPAN
Ventspils
Ionava
Grodno
Novgorod
Cherepovets
Rovno
Zlodin
Dorogobuzh
Tula
Moscow
Shchekino
Novomoskovsk
Voskresensk
Cherkassy
Dneprodzerzhinsk
Odessa
Zaporozhye
Avdeyevka
Lisichansk
Pervaya
Severodonetsk
Dzerzhinsk
Gorlovka
Rossosh
Novolipetsk
Kirovochepetsk
Berezniki
Gubakha
Perm
Nizhni Tagil
Togliatti
Saratov
Ufa
Kuibyshev
Salavat
Magnitogorsk
Khalikovo
Nevinnomysk
Groznyi
Rustavi
Kirovakan
Derbent
Sumgait
Baku
Kemerovo
Novokemerovo
Krasnoyarsk
Tayshet
Angarsk
Kamensk
Komsomolsk-na-Amure
Navoi
Chirchik
Chernorechenskoye
Vakhsh
Fergana
A
B
C
D
E
F
G
H
J
K
Toyo Plants
Creusot Loire Plants
Chemico Plants
USSR Plants (Chirchik)
9th & 10th FYP Programme
Ammonia pipeline planned
Ammonia pipeline under construction
A – Lithuania, Estonia, Byelorussia, Ukraine, Moldavia
B – Central and Central Black Earth Economic Region
C – North-Western Economic Region
D – Siberia, Far East
E – North Caucasus Economic Region
F – Volga Economic Region
G – Volgo-Vyatsk Economic Region
H – Urals Economic Region
J – Georgia, Azerbaidzhan, Armenia
K – Kazakhstan, Uzbekistan, Turkmenistan, Kirgizia, Tadzhikistan

its completion could extend well into the mid-1980s.

Start-up date	Location	Builder	Process	On the map located in the region
May 1973	Nevinnomysk	Toyo	Kellogg	E
April 1974	Severodonetsk	Toyo	Kellogg	A
November 1974	Novomoskovsk	Toyo	Kellogg	B
June 1975	Novgorod	Toyo	Kellogg	C
July 1975	Novomoskovsk	Toyo	Kellogg	B
January 1976	Severodonetsk	Toyo	Kellogg	A
July 1976	Berezniki	Toyo	Kellogg	H
December 1976	Nevinnomysk	Toyo	Kellogg	E
July 1977	Novokemerovo	Toyo	Kellogg	D
July 1977	Togliatti	Toyo	Kellogg	F
July 1977	Ionava	Toyo	Kellogg	A
July 1977	Togliatti	Chemico	Chemico	F
January 1978	Chernorechenskoye	Toyo	Kellogg	K
January 1978	Chirchik	USSR	USSR	K
July 1978	Togliatti	Chemico	Chemico	F
July 1978	Novomoskovsk	Toyo	Kellogg	B
January 1979	Odessa	Creusot-Loire	Kellogg	A
January 1979	Odessa	Creusot-Loire	Kellogg	A
January 1979	Gorlovka	Creusot-Loire	Kellogg	A
January 1979	Gorlovka	Creusot-Loire	Kellogg	A
July 1979	Cherepovets	Toyo	Kellogg	C
July 1979	Dorogobuzh	Toyo	Kellogg	B
July 1979	Togliatti	Chemico	Chemico	F
July 1979	Togliatti	Chemico	Chemico	F
September 1979	Dneprodzerzhinsk	Toyo	Kellogg	A
September 1979	Grodno	Toyo	Kellogg	A
September 1979	Novgorod	Toyo	Kellogg	C
July 1980	Cherkassy	Toyo	Kellogg	A
December 1980	Rossosh	Toyo	Kellogg	B
July 1981	Berezniki	Toyo	USSR	H
July 1981	Kemerovo	Toyo	USSR	D
July 1981	Kemerovo	Toyo	USSR	D
July 1981	Kirovochepetsk	Toyo	USSR	G
July 1981	Rossosh	Toyo	USSR	B
July 1981	Salavat	Toyo	USSR	F
July 1981	Togliatti	Toyo	USSR	F
July 1981	Togliatti	Toyo	USSR	F
July 1981	Togliatti	Toyo	USSR	F

Start-up date	Location	Builder	Process	On the map located in the region
July 1981	Fergana	Toyo	USSR	
July 1981	Perm	Toyo	Kellogg	H
	Fergana	Toyo	USSR	K
January 1982	Angarsk	Toyo	Kellogg	D

B USSR: Occidental Petroleum contract

(i) *Commodity flows involved*

The Occidental:USSR negotiations were undertaken in 1973 and the contract was signed in 1974. According to this contract, Occidental's wholly owned subsidiary, Hooker Chemical Corporation, will deliver to the USSR superphosphoric acid in return for ammonia, urea and potash.

The first deliveries of Soviet ammonia started according to schedule in January 1978 and the first shipments of US superphosphoric acid will take place in 1979. The deliveries will continue until 1997.

	1978	1979	1980	1981-1997
Hooker deliveries of superphosphoric acid in 1,000 tons	10	480	1,000	1,000
USSR deliveries in 1,000 tons				
Ammonia	350	1,200	2,100	2,100 to 1988 then 1,500 to 1997
Urea	250	250	1,100	1,000
Potash	380	830	1,000	1,000

The 1 million tons level of superphosphoric acid delivered to the USSR which will be reached by 1980 is to continue throughout the 20-year period to 1997. On the other hand, the USSR will deliver 2.1 million tons of ammonia pa for the first ten years to 1988 (significantly less during the first two years) and after that the shipments are to decline to 1.5 million tons pa for the rest of the period to 1997. The levels of Soviet urea and potash deliveries will be maintained throughout the period having reached the level of 1 million tpa in 1980.

(ii) *Related projects*

Occidental has also entered into a contract for the supply of technology, design, construction and equipment for two ammonia and superphosphoric acid storage facilities which are to be constructed by the USSR at Ventspils and Odessa. The value of this contract is $100 million. Another related contract valued at $300 million for the supply of technology, design and equipment was signed for the construction of an ammonia pipeline from Togliatti on the Volga to the port facilities which are to be located at Odessa on the Black Sea. Originally two such pipelines were planned. The other was intended to link Togliatti with Ventspils on the Baltic Sea where ammonia storage and port facilities are constructed. However, the latest developments indicate that this project has been postponed indefinitely and ammonia is shipped to Ventspils by rail.

In January 1976 Occidental's subsidiary, Adtek, which is responsible for the construction of harbour facilities, awarded a contract to Honeywell for the supply of process control equipment and refrigeration and storage equipment.

In all, the USSR is thought to be investing over $1,000 million in Western equipment (including $400 million of US technology).

In February 1977 work started on the construction of the first section of the Odessa-Togliatti pipeline between Gorlovka, Panyutino and Odessa. This first stage is constructed by the French company, Entrepose, and the work involves construction of a 14 inch main line to a branch junction 600 km from Odessa; from there a smaller 10 inch line will be laid to Gorlovka which is about 300 km further on. The second stage of this project will see the construction of a 14 inch pipeline from Gorlovka to Togliatti. The completed line will be about 2,100 km long. Also involved in the project are William Bros. who supply some of the equipment.

Another major feature of this massive development is the construction of a fleet of seven refrigerated tankers by the USSR. The first of these, *Yurmala*, was launched in December 1975. It has a capacity of 12,000 cu metres of liquefied ammonia. In order to meet the demands of this ammonia trade the USSR established a whole new shipping company, the Latvian Sea Shipping Co., which will operate these tankers.

(iii) *Pricing and other information*

The terms of the deal guarantee that the USSR will not incur a net loss of foreign exchange. The prices of the US superphosphoric acid will be based on prevailing world market prices and Occidental guarantees that the total value of the acid will not exceed that of the Soviet ammonia, urea and potash. Occidental is to take possession of these Soviet materials on an f.o.b. basis, and it will receive purchase discounts varying between 2.5 and 3 per cent for selling Soviet products.

In November 1978 Occidental won substantial concessions from the USSR on the issue of the price of Soviet ammonia marketed in the United States. The move was necessary in order for the company to compete successfully with other suppliers to the US market such as Mexico and Trinidad. This means that the cost came down from $110-115/ton c.i.f. to $90/ton c.i.f. The higher price established in a contract between Occidental and the USSR in 1976 meant that the company was incurring substantial losses as the US price of ammonia came down.

There is considerable concern in the West as to the effect this trade will have upon established markets (see chapter 5). Last year the total world trade of ammonia was only about 3.2 million tons. This means that the flood of Soviet ammonia onto the world market, generated by this and other contracts, will prove to be a relatively large amount to be absorbed. So far Occidental has only one US customer committed to a purchase of Soviet ammonia – Royster Co. It seems that Occidental will have fewer interested customers than initially foreseen and it will encounter considerable difficulties in marketing the Soviet produced commodities.

Index